I dedicate this book to all animal-loving children. As a child, I was worried about losing my teeth and also nervous whenever I had to go to the dentist. I remember distracting myself by reading lots of science books and learning many amazing things about animals. For example, I loved dinosaurs and discovered that they lost their teeth, too! All these interesting facts made me feel stronger and my worries began to fade away. I hope this book helps you find your passion for the weird and wonderful mouths of the animal kingdom, as well as your own!

Dr Letizia Diamante

What on Earth Books is an imprint of What on Earth Publishing
The Black Barn, Wickhurst Farm, Leigh, Tonbridge, Kent, UK, TN11 8PS
30 Ridge Road Unit B, Greenbelt, Maryland, 20770, United States

First published in the United Kingdom in 2024

Text copyright ©2024 Dr Letizia Diamante
Illustrations copyright ©2024 Ed J. Brown

All rights reserved. No part of this publication may be reproduced or transmitted in any form or by any means, electronic or mechanical, including photocopying, recording, or any information storage or retrieval system, without permission in writing from the publishers. Requests for permission to make copies of any part of this work should be directed to info@whatonearthbooks.com.

Written by Dr Letizia Diamante
Illustrated by Ed J. Brown

Letizia Diamante has asserted her right to be identified as author of this work and Ed J. Brown has asserted his right to be identified as illustrator under the Copyright, Designs and Patents Act 1988.

Staff for this book: Nancy Feresten, Managing Director; Natalie Bellos, Publisher;
Katy Lennon, Senior Editor; Andy Forshaw, Art Director

With thanks to Patrick Skipworth and Richard Pallardy
Trademark notice: Velcro, mentioned on page 38, is a trademark of VELCRO Brand

A CIP catalogue record for this book is available from the British Library

ISBN: 9781804661352

Printer Code: DC/Foshan, China/02/2024
Printed in China

1 3 5 7 9 10 8 6 4 2

whatonearthbooks.com

CONTENTS

A Tasty Introduction 4
Pearly Whites 6
Attack! 10
Losing Teeth 14
Chew it Over 18
Down in One 22
Tremendous Tusks 26
Grow & Gnaw 30
Terrific Tongues 34
Longest Tongue 38
Surprising Spit 42
Bills & Beaks 46
Hidden Talents 50
Mighty Mouths 54
Gigantic Grins 58
Glossary 62
Index 63
Selected Sources 64

A TASTY INTRODUCTION

It's mealtime on the African savanna. A go-away bird pecks at the seed pods on an acacia tree. On the grassland below, a pack of lionesses bare their teeth, ready to pounce on their prey. A wildebeest has been grazing grass for hours, but still wants more. And watch out for that mosquito: it's just about to sip a blood smoothie. Sluuurp!

All these animals have different types of mouths – and these differences are essential to the animals' survival. Their mouths and teeth allow them to access food. A lions' sharp teeth can tear up meat and a bird's strong beak can crunch up seed pods. A wildebeest has a short, flat muzzle and a wide row of teeth, which means it can feed on the short grass of the savanna. A mosquito uses its needle-like mouth parts to pierce the skin of an animal and suck its blood.

But animals use their mouths for so much more than just eating! Some use their mouths for climbing, or for storing food, feeding their babies or scaring away enemies. Get ready to chow down on some fang-tastic animal facts and challenge yourself (and your friends and family) with some guessing games about the most awesome jaws, teeth, beaks, tongues, spit and more in the animal kingdom.

fascicle

PEARLY WHITES

Some teeth are pointy, some are flat, some are small and some are as big as bananas! Teeth allow animals to bite, fight, tear, chew, nibble, grind... and scare off enemies. Each type of toothed animal has a unique set of gnashers.

SPOT THE DIFFERENCE
MILK TEETH vs ADULT TEETH

We have two sets of teeth over the course of our lives. Milk teeth usually start coming up through our gums when we are around six months old. They keep coming through until we have a set of 20. As we grow, our jaws develop and create space for 28 larger and stronger adult teeth. These adult teeth grow after our milk teeth become wobbly and fall out. Some adults also have wisdom teeth. These are extra molars and they can sometimes stay hidden in the gums.

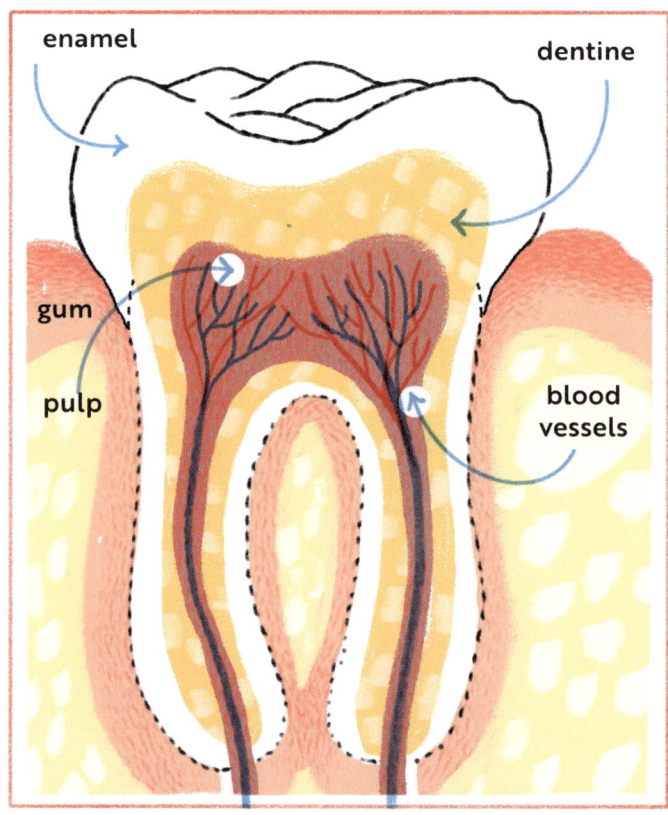

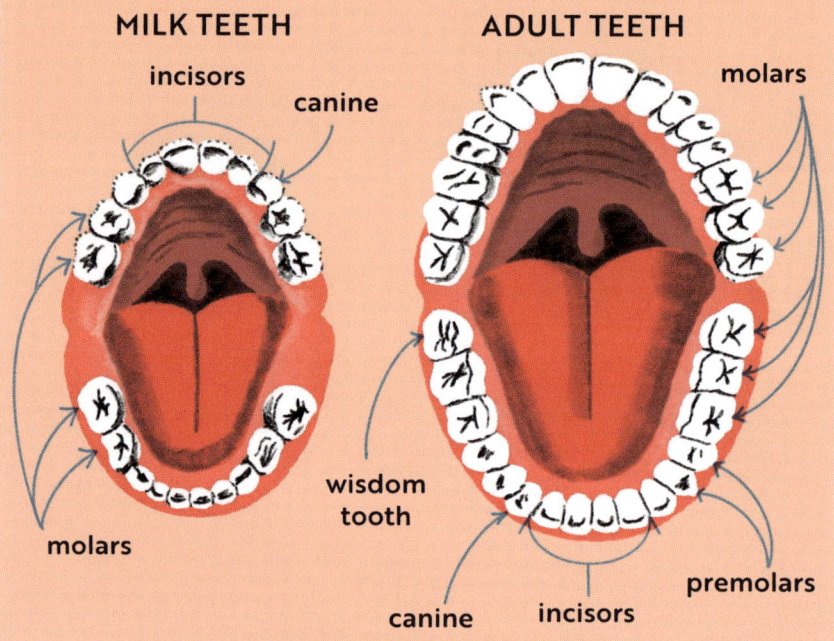

THE INSIDE STORY OF HUMAN TEETH

Teeth are covered in a shiny layer of enamel, which is the hardest substance in the human body. Below the enamel is a layer of dentine, which is also hard, but has microscopic holes in it. Inside the dentine is the pulp, which houses the nerves that allow teeth to feel heat, cold and pain. Blood vessels are also in the pulp. They carry blood to and from the teeth. Surrounding the root of the tooth is a layer of cementum. You cannot see it because it is below the gumline, but it helps 'cement' each tooth to its socket.

Not for the Squeamish

Your mouth is home to billions of bacteria of around 700 different species. Some are healthy, but others are nasty. Bad breath is usually caused by harmful bacteria and bits of food. To avoid it, floss between your teeth and brush for at least two minutes twice or more a day.

MAMMALIAN MOUTHS

The teeth of mammals are so distinctive that experts can often identify a species (a group of similar animals that can breed together) from just a single tooth.

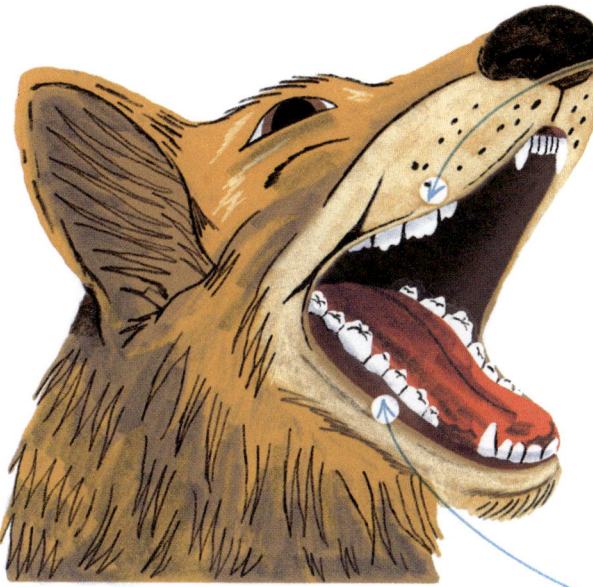

upper carnassials

Meat-eating animals, such as dogs, wolves, cats, bears and weasels, have sets of molars and premolars called carnassials (or shearing teeth). Carnassials have sharp cutting edges and when the animal brings its jaws together, these pairs of teeth work like scissors to slice through flesh and bone.

lower carnassials

Molars and premolars are flat. They are good for herbivores (plant eaters), such as sheep, who need to crush and grind grass so they can digest it.

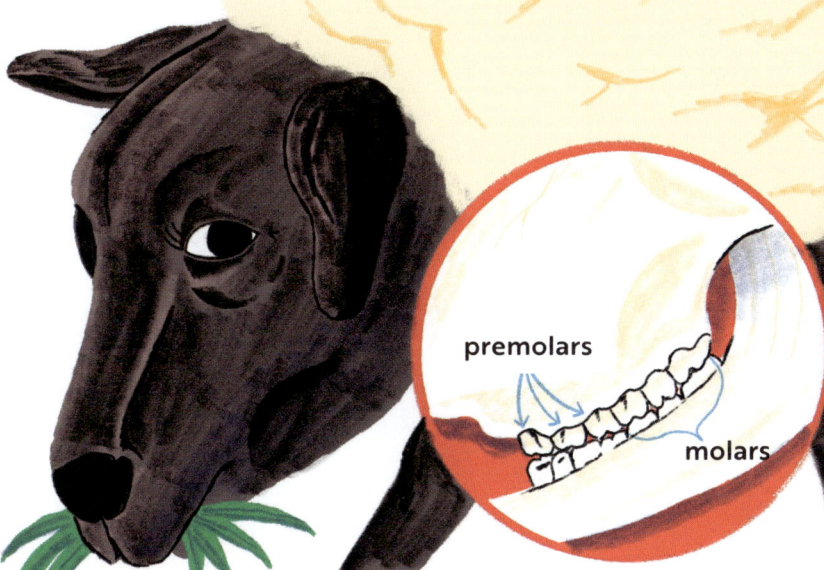

premolars

molars

Incisors (or front teeth) are used for cutting, scooping, picking up objects and grooming. Giraffes have incisors on their bottom jaw and no front teeth on their top jaw. They chew using the molars at the back of their mouths.

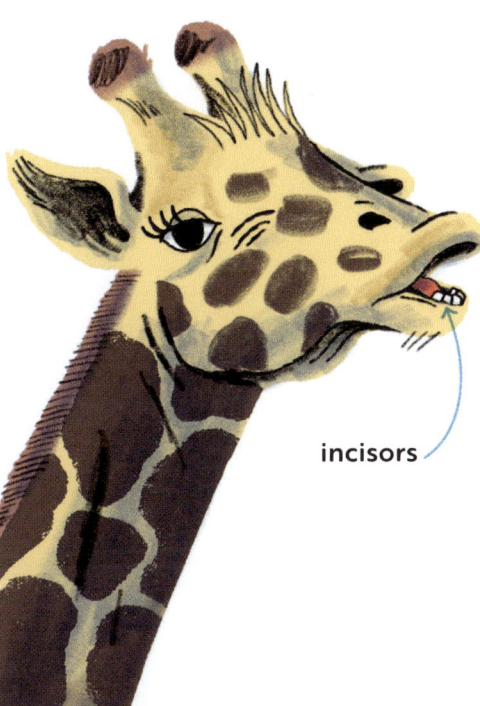

incisors

canines

Pointy canines are great for carnivores (meat eaters), such as wolves, who often need to hold their prey still and tear its flesh to eat it. Now you know: if you come across a suspicious-looking Granny, just like Little Red Riding Hood did, you can tell it's a hungry predator by its teeth!

Snails and slugs are not mammals. They belong to a group of animals called molluscs, which also includes oysters and octopuses. Molluscs are soft-bodied invertebrates (animals without a backbone).

Snails and slugs have a varied diet that includes worms, plants and fungi. But which body part do they use to chew? Turn over to find out!

A tongue-like structure called a radula!

HOW SNAILS EAT THEIR SNACKS

Snails and slugs break up their food by rubbing it with a tongue-like structure called a radula. The radula is covered in thousands of tooth-like denticles that are so small it takes a very strong microscope to see them. As the denticles wear down, new ones grow in their place.

SNAILS IN SNACK-TION

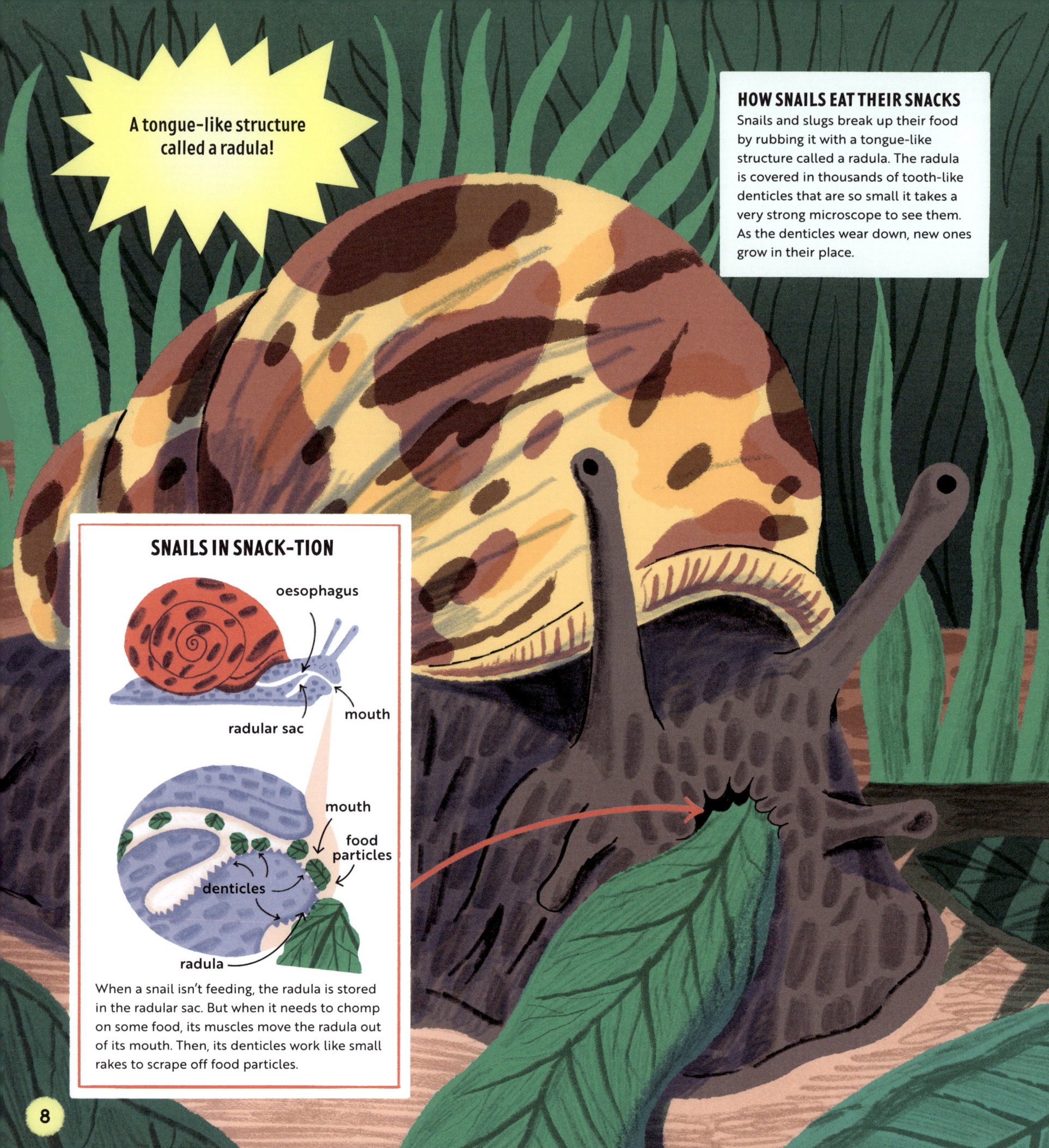

oesophagus
mouth
radular sac

mouth
food particles
denticles
radula

When a snail isn't feeding, the radula is stored in the radular sac. But when it needs to chomp on some food, its muscles move the radula out of its mouth. Then, its denticles work like small rakes to scrape off food particles.

NO TEETH, NO PROBLEM!

Several animal species, such as platypuses, pangolins, birds, turtles and tortoises, are toothless. Let's go to Australia to have a closer look at the platypus – a rather unusual, carnivorous mammal with a special bill. Young platypuses have a set of premolar and molar teeth at the back of their bills, but these fall out and are replaced by grinding pads. Underwater, platypuses search for food and store it in their cheek pouches. When they reach the surface, they grind up the food using their pads.

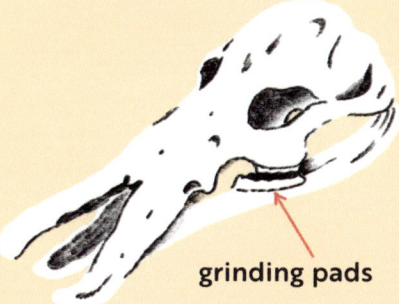

GRIND AND GROW

grinding pads

Unlike most mammal teeth, a platypus's grinding pads are always growing. If they did not continue to grow, all the grinding would wear them down to nothing.

Platypuses close their eyes and ears to swim, but their soft and flexible bills work like prey-detectors and can sense tiny movements of potential food, such as shellfish, insects and worms.

SEA SPONGES

NO MOUTH, NO PROBLEM!

Sea sponges don't have mouths. Instead, they suck seawater in through small holes in their bodies. As the water flows in, it brings in tiny creatures, such as plankton, that the sponges feed on.

Giant tube worms are born with a mouth that seals up as they grow. They absorb sulphur-containing chemicals from the water using their red, feathery-looking plumes. Bacteria inside the tube worms turn these chemicals into energy for both the bacteria and the worm.

GIANT TUBE WORMS

ATTACK!

Meet some of the fiercest, strongest and most majestic predators on Earth! These animals are adapted to attack, capture and devour their prey. Many of them can even ambush creatures bigger than themselves. Often this success relies on the predator's powerful mouth.

DIVE

After spotting its prey, often another bird, a peregrine falcon swoops into a steep dive. It is the fastest hunter in the sky – in fact, the fastest animal on Earth. It can reach speeds of around 300 kilometres per hour. Once it has caught its prey in mid-air, the falcon will use its sharp beak to tear the prey into pieces.

SPOT THE DIFFERENCE
FALCON vs HAWK

FALCON'S BEAK — tomial tooth

HAWK'S BEAK

A hawk's upper beak has a smoothly curved bottom edge, while a falcon's has a small ridge on it. This ridge is called a tomial tooth, but it is not a real tooth. Hawks and falcons kill prey with their powerful claws, but falcons also use their tomial tooth like a wrench – to twist the prey's neck and break its spine.

AMBUSH

Tigers usually hunt in the dark. They stalk their prey, sneaking up as close as possible by crouching on the ground and hiding in the undergrowth. Their hunting strategy is based on surprise. When the moment is right, the tiger pounces on its prey and grabs it with its teeth and claws. With a snap of its jaw, the tiger can break the neck of smaller prey or bite the throat of larger prey. Then it's dinner time!

Tigers are apex predators, which means that they are at the top of the food chain and no other animals prey upon them. They get their choice of things to eat, and their diet includes antelopes, deer, wild pigs, water buffaloes, leopards and crocodiles. They are not picky eaters, and if large animals are not available, they will dine on smaller ones, such as hares, monkeys and peacocks.

How many teeth does a tiger have? Turn the page, if you dare...

11

Thirty! All adult cats have 30 teeth – 12 incisors for cutting, 4 canines for stabbing and ripping, and 10 premolars and 4 molars for crunching and crushing.

INCISOR

PREMOLAR

CANINE

MOLAR

A tiger's tongue is very similar to a house cat's. Find out more on page 37!

A tiger's canine teeth can grow up to 7.6 centimetres long. These teeth have pressure-sensing nerves in them so the tiger can feel the best place to sever the neck of its prey.

Can you spot the gap between this tiger's back teeth and canines? A tiger uses that space in its jaw to hold its prey tightly during an attack.

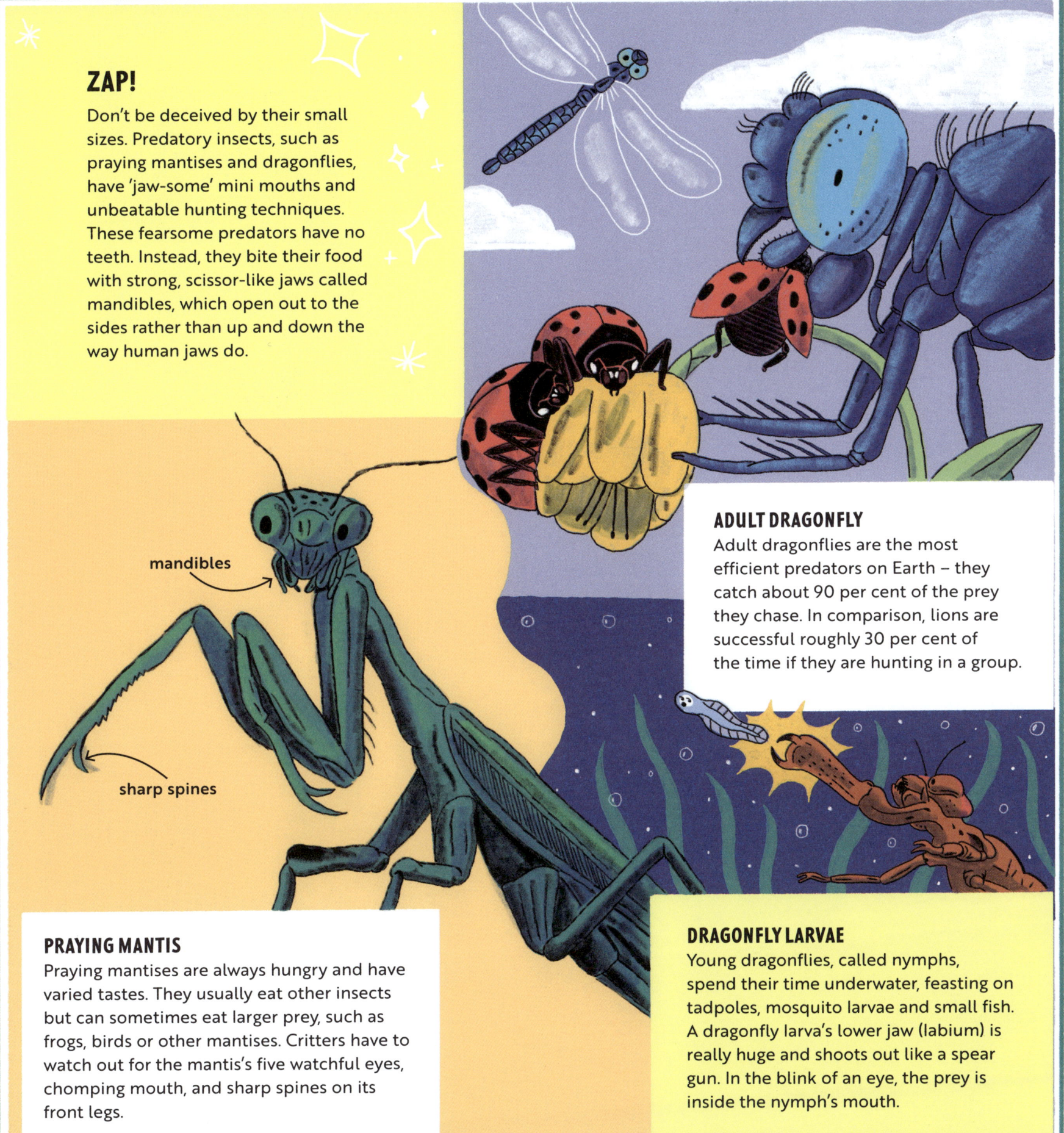

ZAP!

Don't be deceived by their small sizes. Predatory insects, such as praying mantises and dragonflies, have 'jaw-some' mini mouths and unbeatable hunting techniques. These fearsome predators have no teeth. Instead, they bite their food with strong, scissor-like jaws called mandibles, which open out to the sides rather than up and down the way human jaws do.

mandibles

sharp spines

ADULT DRAGONFLY

Adult dragonflies are the most efficient predators on Earth – they catch about 90 per cent of the prey they chase. In comparison, lions are successful roughly 30 per cent of the time if they are hunting in a group.

PRAYING MANTIS

Praying mantises are always hungry and have varied tastes. They usually eat other insects but can sometimes eat larger prey, such as frogs, birds or other mantises. Critters have to watch out for the mantis's five watchful eyes, chomping mouth, and sharp spines on its front legs.

DRAGONFLY LARVAE

Young dragonflies, called nymphs, spend their time underwater, feasting on tadpoles, mosquito larvae and small fish. A dragonfly larva's lower jaw (labium) is really huge and shoots out like a spear gun. In the blink of an eye, the prey is inside the nymph's mouth.

13

LOSING TEETH

We can't regrow our adult teeth, which is why we have to look after them. If you were to accidentally knock out an adult tooth, and your dentist couldn't fix it, you'd be without that tooth forever! Some animals don't have this problem. Whenever crocodiles or sharks wear out their chompers, new ones grow in their place.

Crocodiles have between 60 and 110 teeth, which are replaced continuously. These reptiles can go through 8,000 teeth over a lifetime. Nile crocodiles have the strongest bite in the world – it's 25 to 30 times stronger than the bite of a human! However, they can also be gentle and use their jaws to pick up their babies and carry them around in their mouths.

SPOT THE DIFFERENCE
CROCODILE vs ALLIGATOR

CROCODILE FROM ABOVE

ALLIGATOR FROM ABOVE

Can you tell a crocodile from an alligator? One giveaway is that crocodiles have a V-shaped snout, while alligators have a U-shaped snout. Also, crocodiles have some teeth that poke out and can be seen even when their mouths are closed.

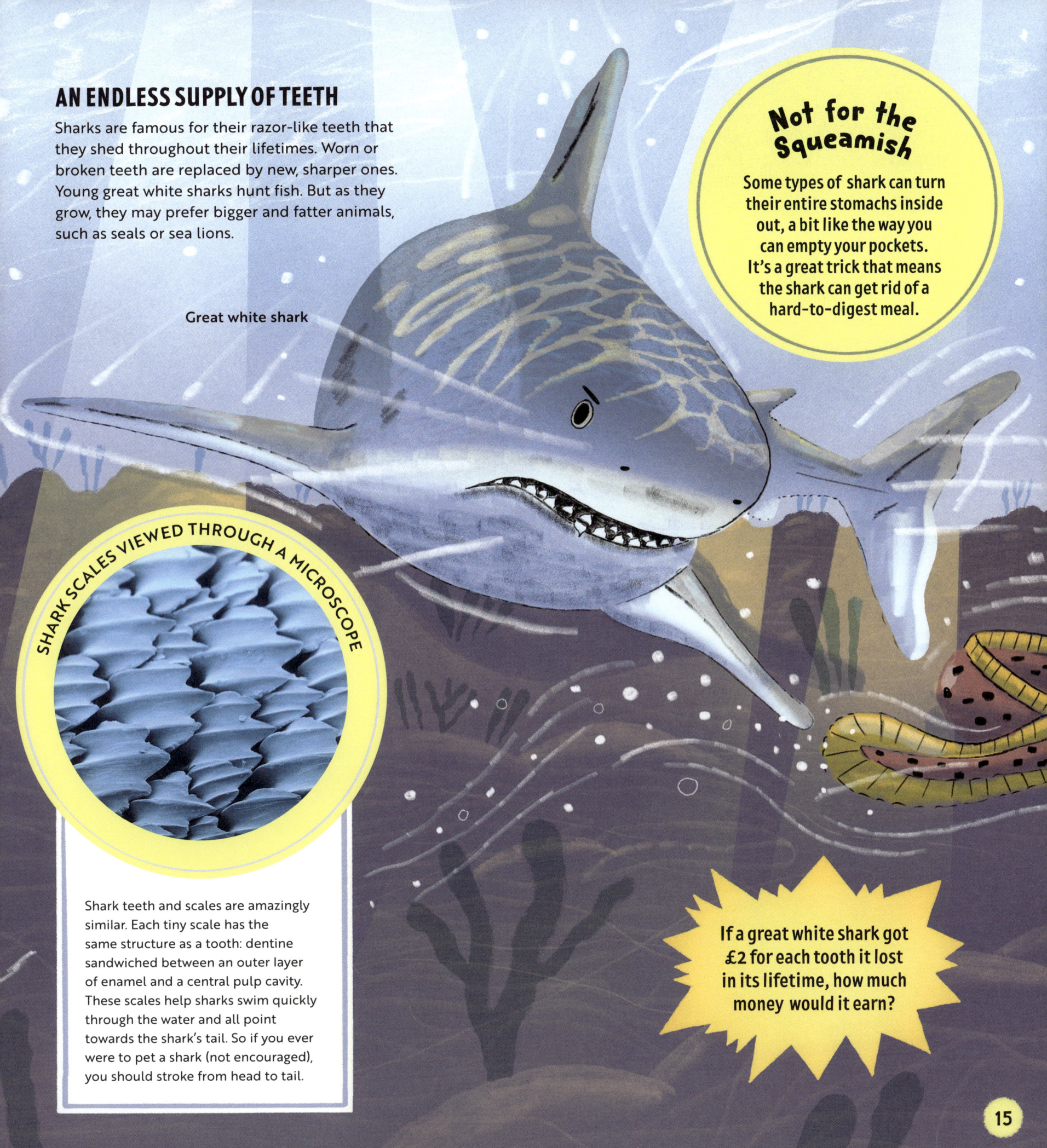

AN ENDLESS SUPPLY OF TEETH

Sharks are famous for their razor-like teeth that they shed throughout their lifetimes. Worn or broken teeth are replaced by new, sharper ones. Young great white sharks hunt fish. But as they grow, they may prefer bigger and fatter animals, such as seals or sea lions.

Great white shark

Not for the Squeamish

Some types of shark can turn their entire stomachs inside out, a bit like the way you can empty your pockets. It's a great trick that means the shark can get rid of a hard-to-digest meal.

SHARK SCALES VIEWED THROUGH A MICROSCOPE

Shark teeth and scales are amazingly similar. Each tiny scale has the same structure as a tooth: dentine sandwiched between an outer layer of enamel and a central pulp cavity. These scales help sharks swim quickly through the water and all point towards the shark's tail. So if you ever were to pet a shark (not encouraged), you should stroke from head to tail.

If a great white shark got £2 for each tooth it lost in its lifetime, how much money would it earn?

£40,000 or even more! The great white shark can lose more than 20,000 teeth in its lifetime.

A GREY NURSE SHARK

A shark's teeth are not anchored to bony jaws like ours are, so when a tooth is damaged, another can flip forwards into its place.

The teeth of some sharks are covered with fluoride. This is the same mineral that is used in toothpaste, which helps prevent cavities.

FOSSIL ADVENTURE

All dinosaurs had several sets of teeth that they could replace throughout their lives. When a dinosaur lost a tooth, it would sometimes become preserved as a fossil. Dinosaur teeth can tell us a lot about their owners: which species they belonged to; what type of food they ate and whether they chewed or crushed it, or just swallowed their food whole.

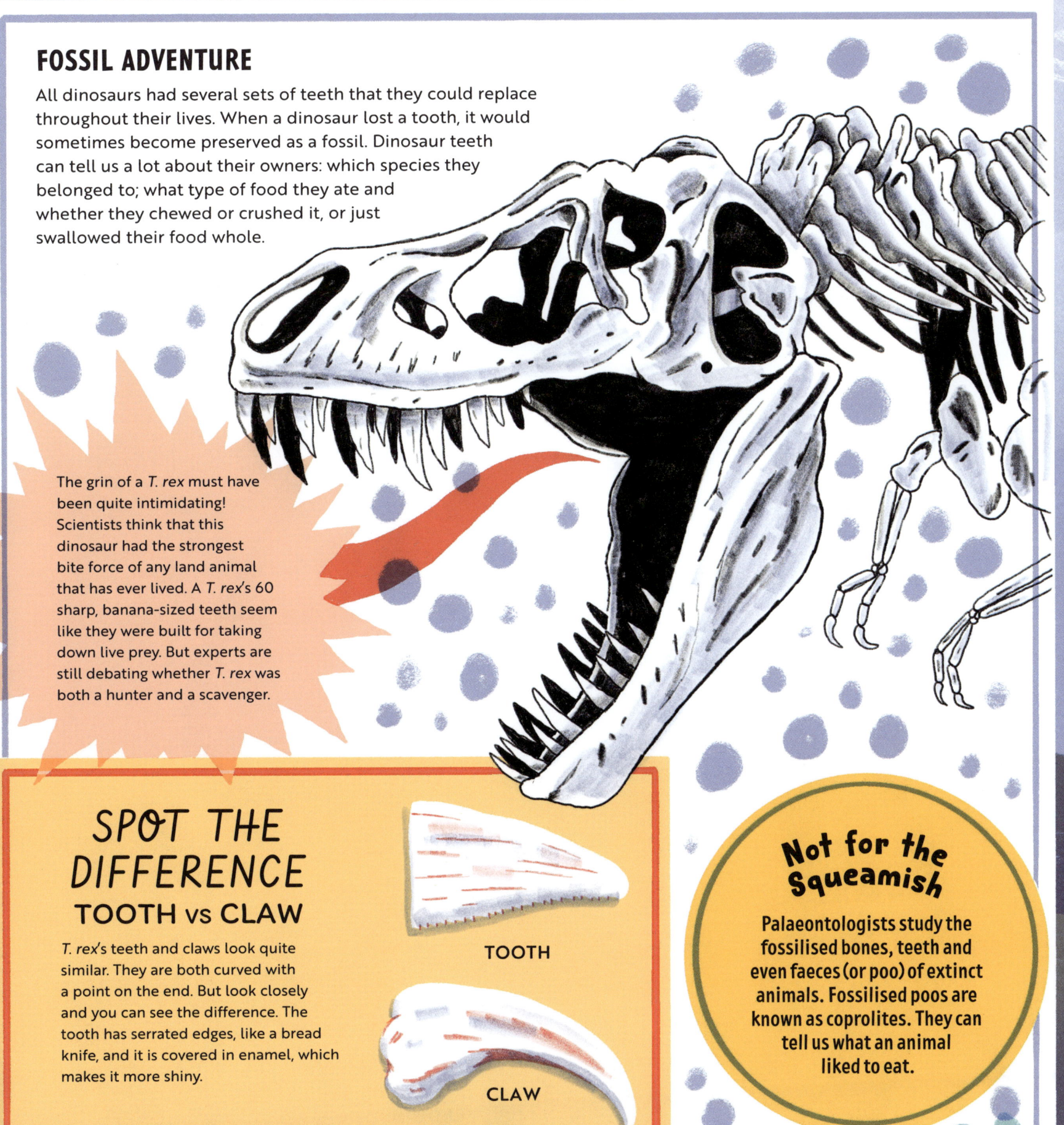

The grin of a *T. rex* must have been quite intimidating! Scientists think that this dinosaur had the strongest bite force of any land animal that has ever lived. A *T. rex*'s 60 sharp, banana-sized teeth seem like they were built for taking down live prey. But experts are still debating whether *T. rex* was both a hunter and a scavenger.

SPOT THE DIFFERENCE
TOOTH vs CLAW

T. rex's teeth and claws look quite similar. They are both curved with a point on the end. But look closely and you can see the difference. The tooth has serrated edges, like a bread knife, and it is covered in enamel, which makes it more shiny.

TOOTH

CLAW

Not for the Squeamish

Palaeontologists study the fossilised bones, teeth and even faeces (or poo) of extinct animals. Fossilised poos are known as coprolites. They can tell us what an animal liked to eat.

CHEW IT OVER

Many plant-eating animals spend their days grazing, browsing, chewing and digesting their food. They need to eat a large quantity of plants and chew them a lot in order to get all the nutrients they need. Cows are chomping champions and spend at least eight hours a day chewing. Other herbivores also munch all day. They might go for a grassy feast, juicy fruits, leafy treats or even prickly cacti!

SPIKY SNACKS

Camels can eat spiny plants that you wouldn't even touch. They grind their food by moving their mouths in a figure-of-eight shape. The inside of camels' cheeks are coated with hard structures called papillae, which guide the spines straight down the throat.

TO BURP OR NOT TO BURP?

Horses browse and graze for up to 18 hours a day. They chew their food and swallow it. They have a one-way throat, meaning that they cannot vomit or burp. Other herbivores, such as cows, giraffes, sheep, gazelle and deer, have a different way of processing their food: they chew it, swallow it, bring it back up, re-chew it and then re-swallow it. This process helps them get the most nutrients out of everything they eat.

Not for the Squeamish

Cows have four stomachs. Bacteria in the first stomach break down the food using a process that produces a lot of gas, especially methane. This gas then makes its way out mainly as burps (and also a bit as farts)! We have a different digestive system, and our burps are mostly caused by extra air that we swallow while drinking and eating.

Grazers and browsers have a variety of techniques for snipping off the best plant parts. Some, like horses, have a full set of incisors. They use them, along with their lips and tongues, to grasp their favourite greens. Others, like cows and sheep, rely more heavily on their tongues and lips because they do not have incisors on their top jaws.

Horses' teeth can tell us if one horse is older than another. Can you guess how?

One way is by the colour! The horse on the right has yellowed teeth, which means it is older than the horse on the left. Read on for more ways that horses' teeth can tell us how old they are.

The colour, wear, angle and slope of a horse's front teeth change with age. A young foal starts out with milk teeth but they are lost and replaced by the time the horse is five. A horse's milk teeth start off white but turn yellow over time and its adult teeth are a brownish yellow. By the time it is 11 years old, a horse has ground down its teeth with years of grazing.

SPOT THE DIFFERENCE
YOUNG vs OLD

5 YEARS OLD | 20 YEARS OLD

Another way to tell the age of the horse is to check the angle between its upper and lower incisors. This angle changes from wider to narrower as the horse gets older.

LONG AND GROOVY TEETH

INCISORS — These vertical grooves first appear on a horse's teeth when the horse is around ten years old. The grooves change over time and have usually disappeared by the time the horse is 30.

MOLARS — Horses' teeth are very long and mostly hidden below the gum line. They push out a few millimetres per year, as the teeth wear.

GRAZING AND BROWSING

There are five species of rhino: the greater one-horned rhino, and the white, black, Sumatran and Javan rhinos. They are all herbivores and eat mostly grass, but they also eat leaves and other plant matter. Black rhinos have no front teeth and use mainly their lips to grip food and their cheek teeth to grind it.

Oxpeckers eat ticks and insects from the skin of rhinos. They have also been known to eat skin, blood and saliva.

Black rhino

SPOT THE DIFFERENCE
WHITE vs BLACK RHINOS

Both species of rhinoceros are in fact grey – except when they have been in a mud bath! The white rhino is much longer and bigger than the black one, but there is another important difference. White rhinos (square-lipped rhinos) are grazers: they have very flat and wide lips, which work like lawnmowers. Black rhinos (hook-lipped rhinos) tend to feed on leaves and branches. Their pointed, grasping lips grab hold of trees and shrubs to eat.

WHITE RHINO

BLACK RHINO

DOWN IN ONE

Some animals don't need to chew their food – they have teeth or other mouth parts that are great at gripping, ripping, tearing and shredding food instead. Crocodiles, alligators and caimans either swallow food whole or gobble huge chunks. And birds... well, you'll see.

GRIT IN THE GIZZARD

Some birds swallow pebbles and grit (small stone or sand particles) to help with digestion. The movement of the grit in their gizzard (part of their stomach) grinds the food for them.

NO NEED TO CHEW

Can you imagine eating live, slimy worms? Well, they're some birds' favourite food and the birds don't even need to chew them, they just slurp them up like strands of spaghetti. Penguins love to eat krill, squid and fish. Their mouths and tongues are lined with sharp, backwards-pointing spines that stop their prey from escaping.

Not for the Squeamish

Most parents need to provide food for their babies and penguin parents have an interesting way of doing this. They find, eat and swallow food and then regurgitate it (bring it back up) into their chicks' mouths for them to eat. This process helps soften the food, making it easier for the baby birds to digest. It also makes it a lot harder for another animal to steal the food!

A TIGHT SQUEEZE

Anacondas and caimans are two other animals that eat their prey whole. They are both top predators in the Amazon rainforest and very rarely come head to head, but when they do, they can engage in dramatic fights to the death. Anacondas are slow moving on land, but fast in water. The anaconda's strategy is to coil its long, heavy and muscular body around the caiman and squeeze tight. But it needs to watch out as the caiman has sharp teeth, a strong bite and can deliver powerful slashes with its tail...

Anaconda versus caiman: who will win?

The snake won! It squeezed the caiman tight and then opened wide to swallow it down.

ONE BIG GULP

Snakes swallow their food in one slow gulp! The heaviest snake in existence is the green anaconda. Female green anacondas can weigh as much as 250 kilograms, which is about the same as a male grizzly bear! They also grow to over 6 metres long and measure more than 30 centimetres in diameter. Males are around half the size. Anacondas can open their mouths really wide to swallow their prey in one go. After a big meal, they can go weeks or months without more food.

glottis

Have you ever tried to breathe at the same time as swallowing? Impossible, isn't it? A snake can do this with ease! It can breathe through its mouth while feasting thanks to its extendable glottis: this little tube functions like a snorkel to let in air.

Green anacondas are not venomous, but other snakes, such as vipers, have long sharp teeth (fangs) to inject venom (poisonous saliva) into their prey to disable or kill it. When they're not being used, the fangs fold flat against the inside of the snake's mouth. This clever mechanism stops the snake from accidentally biting itself.

A SNAKE SHOWING OFF ITS FANGS

THE CATAPULT EFFECT

Frogs can also swallow their prey whole, but they need to catch it first. A frog's secret weapon is its slimy tongue. Unlike a human tongue, a frog's tongue is attached to the front of its mouth, so it can use it like a catapult to strike at its prey. Three, two, one, go!

In the blink of an eye, a frog's extra-sticky tongue and saliva catch an insect and drag it inside its mouth.

TOO BIG TO CHEW

One question that's long puzzled scientists is: how did big, plant-eating dinosaurs, such as *Argentinosaurus*, have enough time in the day to eat all the food that they needed? Elephants spend 16 to 18 hours feeding but are 14 times smaller than *Argentinosaurus*. That suggests that there simply weren't enough hours in a day to satisfy the *Argentinosaurus'* appetite. Some scientists believe the only way *Argentinosaurus* could have grown so big was by not wasting time chewing, but instead just gobbling its food down in chunks. Gulp!

The frog closes its eyes and pushes them down into the roof of its mouth. This helps push food down the frog's throat.

TREMENDOUS TUSKS

Some mammals have long teeth called tusks. The tusks usually come in pairs and are constantly growing. They are made of smooth, white dentine – the tough substance found under the enamel of our teeth. Tusks are some of the longest teeth in the animal kingdom and can have many different uses.

WALRUSES' GRAPPLING HOOKS

Both male and female walruses have tusks. These are enlarged upper canine teeth that can grow to be about 1 metre long. Walruses often use their tusks like ice picks, to pull themselves up out of the water and onto the ice. They also use them to punch breathing holes in the ice, to scratch an itch and to fight each other.

UNICORN OF THE SEA

Narwhals – although usually only male narwhals – grow one skewer-like tusk out of their upper lip. Only very rarely will a narwhal grow two tusks at once. The tusk is actually their left canine tooth and can grow to more than 3 metres long. Its purpose is still a bit of a mystery. Scientists think narwhals may use the tusk to attract partners, fight other males, or strike fish to stun them before eating them. They might also use their tusk to 'taste' the amount of salt in the water.

A MALE NARWHAL SURFACING

A narwhal's tusk has an anticlockwise spiral. Narwhals gain a ring on the tusk for each year of growth. Scientists can study narwhals' tusks to reveal changes in their diet and the Arctic environment.

TUSKS AS TOOLS

An elephant's tusks are its front teeth (incisors), which it uses a bit like we use our hands. Just like humans tend to be either right- or left-handed, elephants will have a preferred tusk. Elephants use their tusks to perform lots of tasks, such as breaking branches and peeling bark off trees to eat, and for defence or to attack opponents during fights. Most male and female African elephants have tusks, but only male Asian elephants have them.

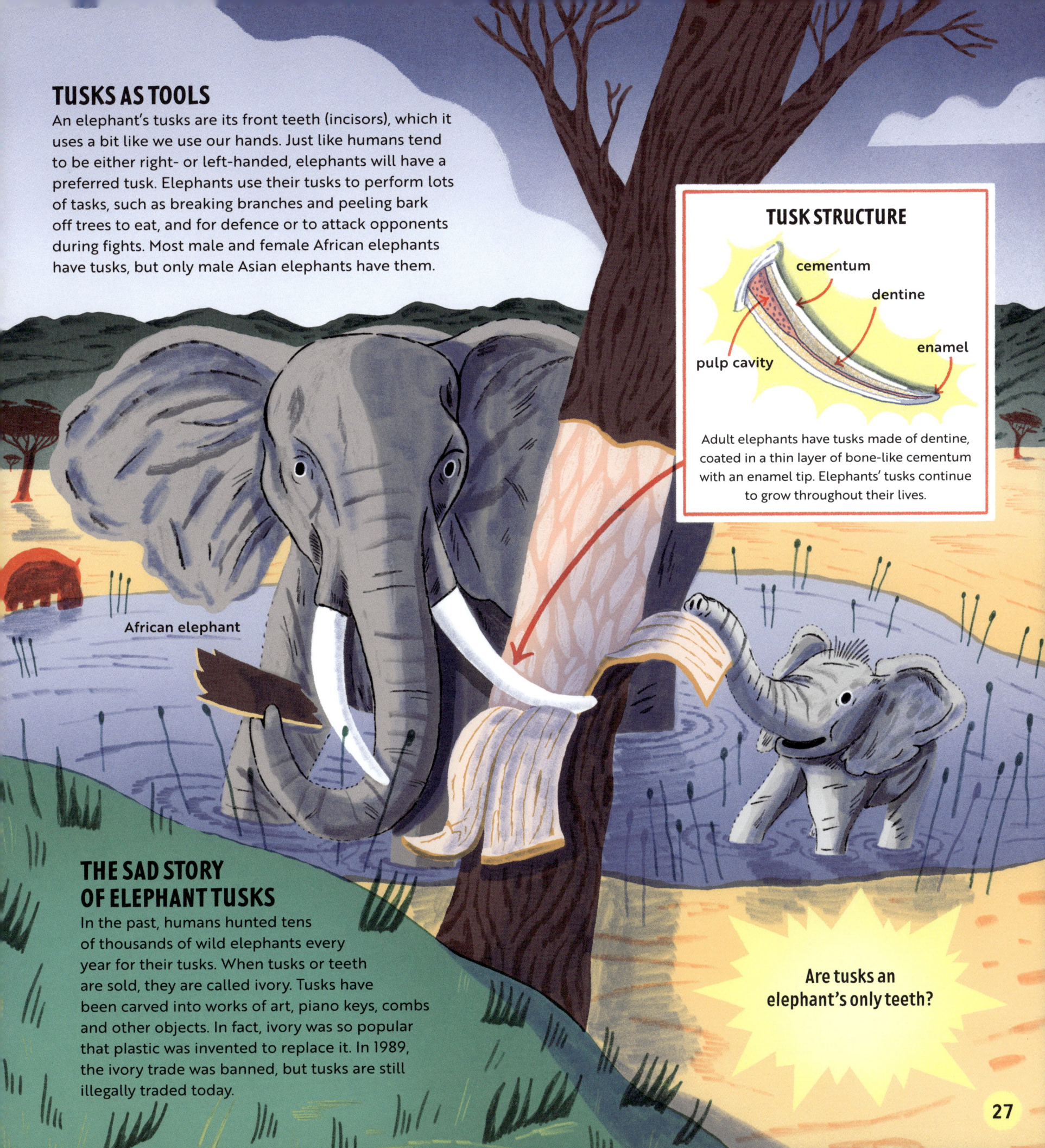

African elephant

TUSK STRUCTURE

cementum
dentine
enamel
pulp cavity

Adult elephants have tusks made of dentine, coated in a thin layer of bone-like cementum with an enamel tip. Elephants' tusks continue to grow throughout their lives.

THE SAD STORY OF ELEPHANT TUSKS

In the past, humans hunted tens of thousands of wild elephants every year for their tusks. When tusks or teeth are sold, they are called ivory. Tusks have been carved into works of art, piano keys, combs and other objects. In fact, ivory was so popular that plastic was invented to replace it. In 1989, the ivory trade was banned, but tusks are still illegally traded today.

Are tusks an elephant's only teeth?

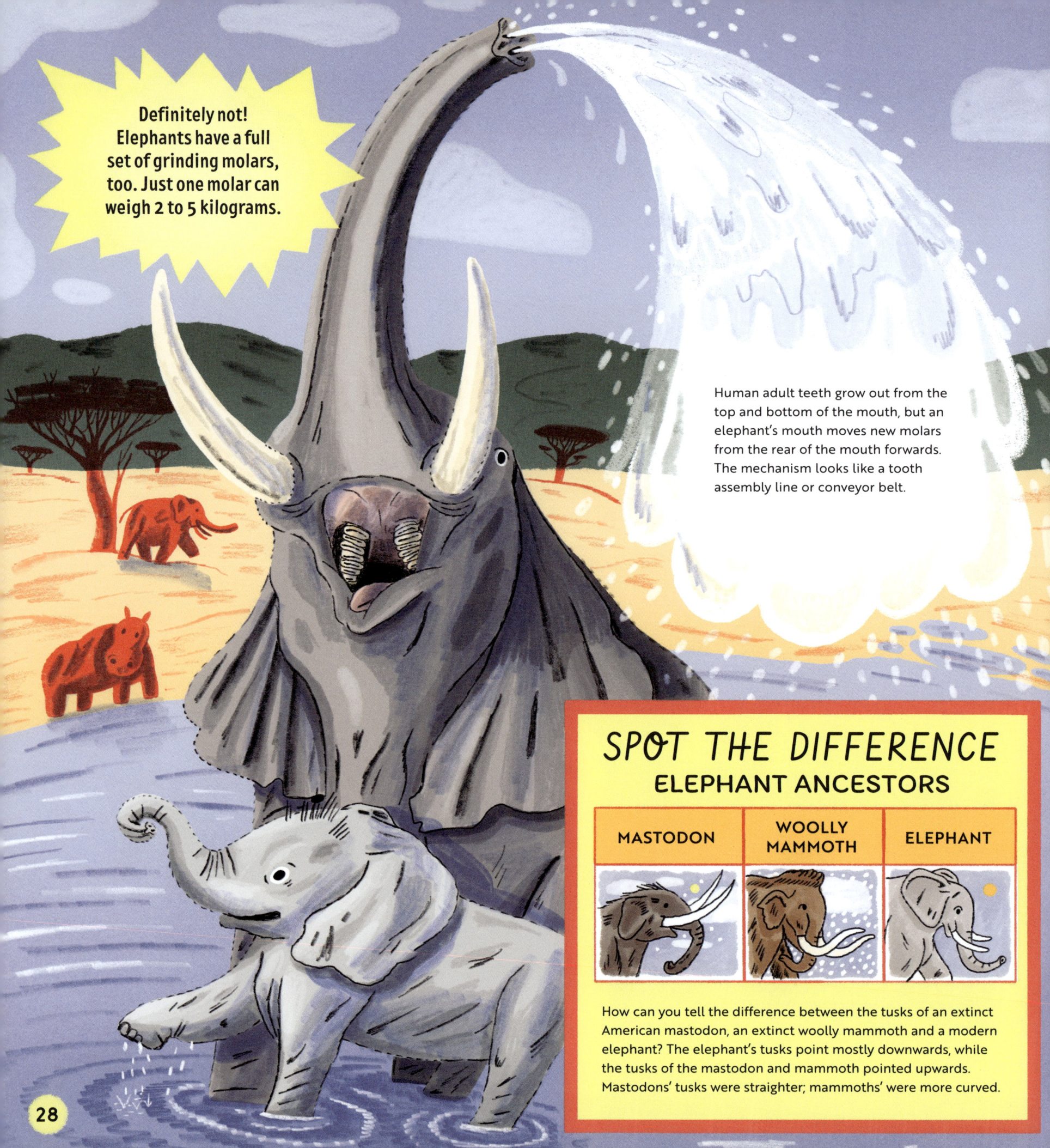

ONE ENORMOUS MOUTH

Although hippopotamuses are mostly herbivores, they have massive tusk-like bottom canine teeth that typically measure almost half a metre. But hippos aren't predators. They use these evil-looking teeth to bite their opponents during fights.

Hippos have the largest mouth of all land animals! They can open their jaws to almost 180 degrees. In a fully grown male hippo, this is about 1.2 metres wide – large enough to fit a golf club inside.

TUSKED DEER

Most deer have antlers, but water deer have tusks instead. They use these overgrown teeth to fight off predators. Males also use them to fight other males for potential mates.

GROW & GNAW

Rodents (such as squirrels, beavers and rats) and rabbits have chisel-like incisors that keep growing throughout their lives. To stop their incisors from growing too long, the animals chew on tough foods to wear them down. Can you imagine being able to crack open a walnut or take down a tree just with your teeth? Let's meet the animals that can.

Squirrel incisors grow at a rate of about 15 centimetres per year.

Squirrels eat lots of different foods, including seeds, fruit, roots and vegetables. They also love nuts! Squirrels collect and bury nuts underground in the summer and autumn, then dig them up in winter, when there's not much fresh food around.

DON'T IG-GNAW IT

Rats gnaw almost everything they come across. They'll munch through wood, cement, rubber, wire and a long list of other materials. To keep the length of their chompers in check, sometimes they grind them. A rat is more inclined to do that when it is relaxed and content, similar to when a cat purrs. At other times, a rat might behave in this way to comfort itself, especially when it is afraid or in pain.

Not for the Squeamish

Rodents can't vomit. This is bad news for rats because it's why rat poison is so effective.

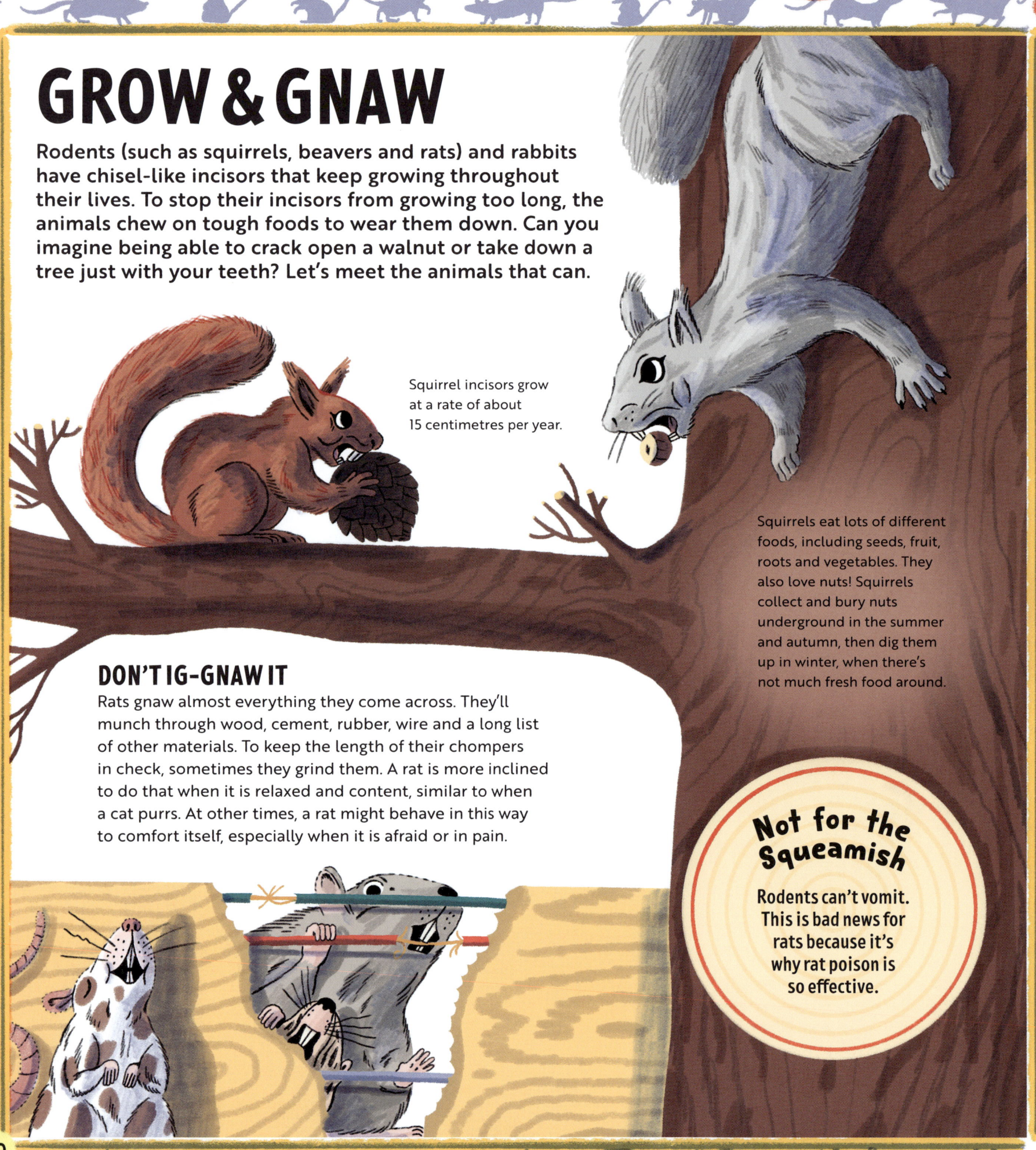

SPEED GNAWERS

You wouldn't ever need an axe if you had teeth like a beaver's. A beaver can swiftly chew through trunks, shrubs and branches, bringing down a 15-centimetre-wide tree in under an hour. Then it drags the cut log to a river to build a dam and a home for itself (called a lodge).

SPOT THE DIFFERENCE
WHO DID THIS?

A lot of animals chew, rub or claw on shrubs and trees, and some leave behind tell-tale marks that let you know who's been there. Beavers cut down trees, leaving a conical stump. Porcupines and squirrels climb trees and gnaw the bark (squirrel bite marks are smaller). Bears scratch tree trunks to mark their territory.

BEAVER **SQUIRREL**

PORCUPINE **BEAR**

Why are beaver teeth orange?

A beaver's teeth (and a porcupine's, too) are made partly of iron, which is a metal that rusts when exposed to air.

SELF-SHARPENING
To chew through wood, beavers' front teeth need to stay super sharp. Instead of getting dull with use like a knife or an axe, a beaver's teeth actually get sharper the more the beaver uses them! Constant chewing wears away the back of the tooth faster than the front, keeping a sharp edge.

ALL A-QUIVER!
Porcupines have sharp spines called quills on their backs. When they're scared, they rattle their quills to warn predators not to get too close. They will also chatter their teeth, just like humans do when they're cold.

TOOTHY TROUBLES

Animals have tooth problems, too. For example, if a rabbit has a tooth that does not wear down at the same rate as its other teeth, it can seem like it's grown a tusk. But it's actually just an extra-long tooth. If that tooth gets too long, the rabbit may be unable to eat and will start to lose weight. Then it's time to call a vet!

HIDDEN TEETH

2nd incisors
1st incisors

Rabbits have four large incisor teeth, two on the top and two on the bottom. They also have a second set of little incisors, which are hidden behind their first incisors. You won't see them unless you look inside a rabbit's mouth.

Not for the Squeamish

Not all the food a rabbit eats is completely digested the first time round. To get the maximum goodness from everything it eats, a rabbit will snack on its poo, too!

SCREAM LIKE A MARMOT!

Another interesting rodent is the marmot. Like porcupines, marmots chatter their teeth to scare off intruders and predators, but marmots have another trick up their sleeve, too: their scream! When a predator approaches, their high-pitched scream tells other marmots to 'Hide! Go back to the dens!' The loud cries can differ from one marmot family to another.

TERRIFIC TONGUES

Stick it out, lick an ice cream, try to touch your nose – is that all you can do with your tongue? This muscular organ can bend into lots of different shapes. Most other muscles in the body develop around a bone, but tongue muscles don't. If an animal (including a member of the human species) pulls faces at you or sticks its tongue out, look carefully. You could discover amazing things!

TONGUE-TASTIC

Did you know we have unique tongue prints as well as unique fingerprints? Everybody has different tongue wrinkles, ridges and other marks. The tongue is also covered with hundreds of small bumps of varying sizes, called papillae, which contain taste buds. We can recognise five tastes: sweet, bitter, salty, sour and umami. Umami is a kind of savoury taste you get when eating foods such as tomato sauce, broth, soy sauce and Parmesan cheese.

papilla
taste bud
papilla

The tongue rarely gets tired, unless you recite a tongue twister. Try the following challenge. Repeat these words as fast as you can, over and over, and try not to get in a tangle: 'terrible teeth, terrific tongue'.

WELCOME TO THE WORLD!

Cows lick their calves clean at birth: it's their way of saying, 'Welcome to the world!' The licking also encourages blood flow and breathing and helps to activate the calf's digestive system. You might also see adult cows licking each other: this behaviour is called social grooming.

TONGUES OF MANY COLOURS

The human tongue is pink, but the animals on this page have tongues in different, far wackier colours! Each tongue has a special 'power' that helps the animal survive.

KOMODO DRAGON
Komodo dragons are the largest living lizards. You can find them in the wild in some parts of Indonesia. They use their tongues to sniff out prey.

POLAR BEAR
Polar bears can get hot running and playing in the sun. They pant with their mouths open to cool themselves off.

BLUE-TONGUED SKINK
Found in Australia, Indonesia and Papua New Guinea, blue-tongued skinks feed on insects and plants. They stick out their tongues to confuse predators when threatened.

GIRAFFE
Giraffes can spend as much as 18 hours a day stretching their long necks to munch on the leaves of the topmost branches of the acacia tree. They use their half-metre-long tongues to dodge the thorns.

PRASINOHAEMA
A pink tongue in humans is a sign of good health. But that's not the case with these lizards...

Can you guess what colour tongue each of these animals has?

Yellow, black, pink and patchy, blue, and aqua!

KOMODO DRAGON
The forked tongue of the komodo dragon is yellow.

GIRAFFE
A giraffe's tongue looks black because it contains lots of melanin, the pigment (coloured protein) that also gives colour to human skin.

POLAR BEAR
A polar bear is born with a pink tongue that develops black or dark patches as it grows.

Not for the Squeamish
The tongue-eating louse (which is actually a type of crustacean) eats the tongue of its fish victim and lives in the tongue's place in the fish's mouth.

BLUE-TONGUED SKINK
The clue was in the name with this one... it's blue, of course!

PRASINOHAEMA
The prasinohaema has an aqua (blue-green) mouth and tongue, as well as green bones and muscles. It is a rare example of an animal with green blood.

GROOMING

Did you know that domestic cats can spend up to half of their waking hours licking and nibbling their fur? Cats – including big cats, such as snow leopards – groom to clean themselves, remove odours and get rid of loose fur. Their rough tongues act like a hairbrush, smoothing out the coat.

THE TINY, HOOKED BARBS ON A CAT'S TONGUE

If a cat licks you, it can feel like sandpaper on your skin because cat tongues have little backward-facing hooks. The hooks help the cat groom itself, which it does by licking its fur.

TASTING

A cat's tongue has around 470 taste buds: that's a really small number compared with up to 10,000 taste buds on a human's tongue and around 1,700 on a dog's tongue. Cats generally enjoy fatty, savoury treats and foods with different textures. They can't taste sugar and hate bitter flavours. Most cats also prefer their food warm – the temperature of freshly killed prey.

DRINKING

Cats have a very clever way of drinking. A cat will touch the surface of the water with the top of its tongue and draw it back up quickly. This causes a little jet of water to fly into the air, which the cat can trap in its mouth by quickly snapping it shut. Cats can lap up water at a speedy rate of four laps per second.

LONGEST TONGUE

Not only do tongues come in various colours, but they also come in many different sizes. Some tongues are teeny while others are loooooooong! And some insects have a tongue-like structure called a proboscis. There are many contenders in the animal kingdom for the title of the longest tongue compared to body size. Can you guess which animal will be the winner? Find the answer at the bottom of page 41.

CONTESTANT #1

THE WOODPECKER

A woodpecker's beak is strong and sturdy, with a chisel-like tip for drilling holes in the wood and bark of trees. However, the most unusual feature of the woodpecker's body is its extraordinary tongue, which is nearly three times longer than the bird's beak.

CONTESTANT #2

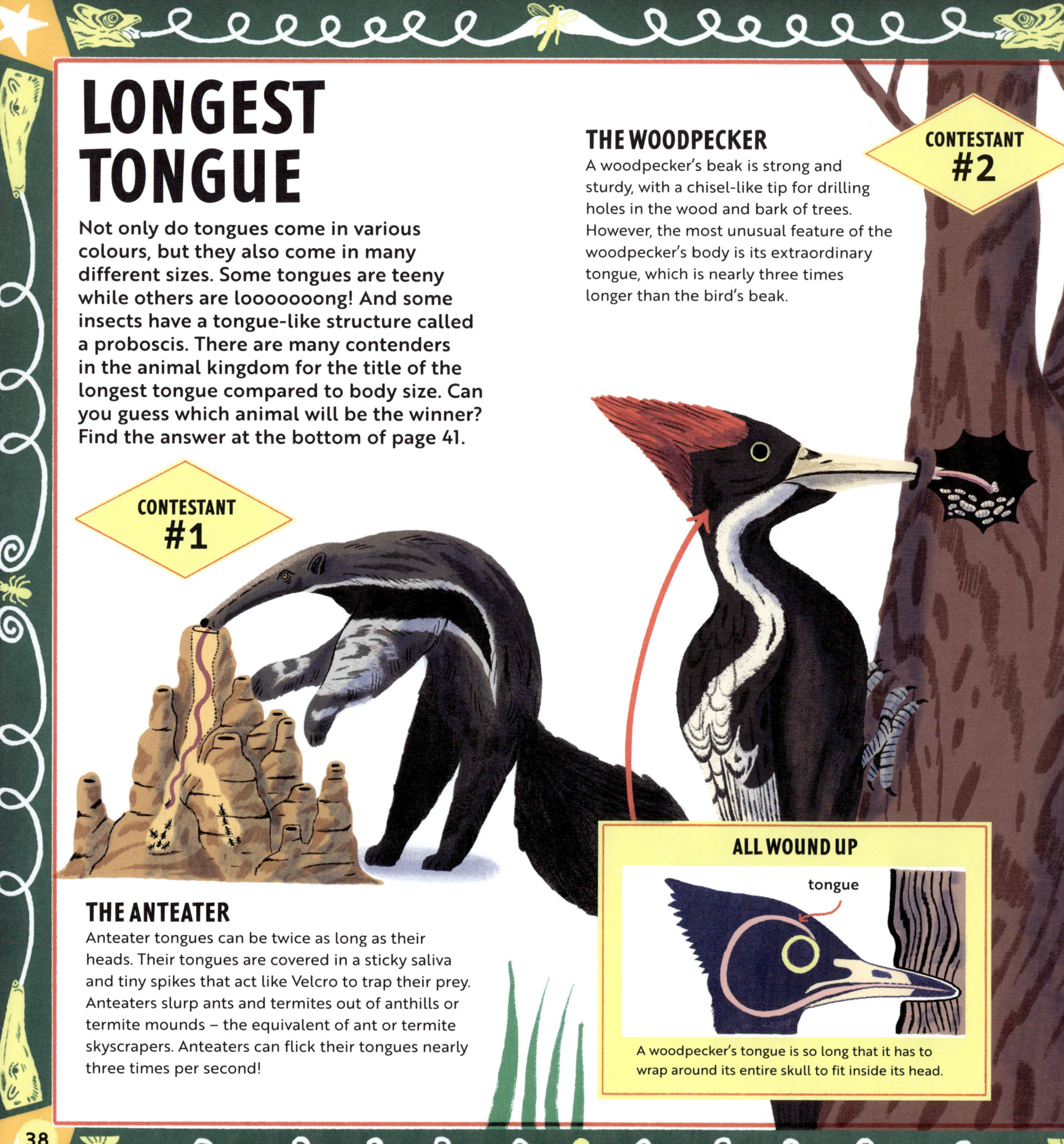

ALL WOUND UP

A woodpecker's tongue is so long that it has to wrap around its entire skull to fit inside its head.

THE ANTEATER

Anteater tongues can be twice as long as their heads. Their tongues are covered in a sticky saliva and tiny spikes that act like Velcro to trap their prey. Anteaters slurp ants and termites out of anthills or termite mounds – the equivalent of ant or termite skyscrapers. Anteaters can flick their tongues nearly three times per second!

THE CHAMELEON

Chameleons have colour-changing skin, eyes that can look in almost any direction and tongues about twice as long as their bodies. A chameleon will often ambush its prey. It will wait until a cockroach, cricket, grasshopper or butterfly ventures close enough, then it shoots out its tongue – the ultimate weapon.

CONTESTANT #3

Panther chameleon

Even when a chameleon is focused on hunting, it will notice you if you try to sneak up on it from behind. This is possible because a chameleon's eyes swivel completely independently of one another. One eye might follow prey while the other looks out for danger and predators.

Chameleons tend to be able to darken or lighten their usual pattern to match their surroundings, but not all of them create particularly vivid colours. The most famous colourful chameleon is the panther chameleon, which is found in the wild only in Madagascar.

A chameleon's tail and toes grip and wrap around branches while the animal hunts, rests and eats.

Will this chameleon manage to catch its grasshopper prey?

Zap! Yes! The chameleon's sticky tongue successfully captured its meal!

A chameleon hunts by hurling out its tongue to catch prey on the sticky tip and then pulling it back into its mouth. The record for fastest tongue belongs to *Rhampholeon spinosus*, also known as the rosette-nosed chameleon. It moves from 0 to about 100 kilometres per hour in one-hundredth of a second: about 200 times faster than a Ferrari SF90 Stradale.

At the tip of a chameleon's tongue, a thick, honey-like substance – nearly 400 times as thick as our saliva – snags prey and brings it back to its mouth.

This tongue can pull in an object that weighs up to about a third of the chameleon's body weight. Can you imagine gobbling up a sweet that weighs as much as a small dog?

THE TUBE-LIPPED NECTAR BAT

The tube-lipped nectar bat can extend its tongue one-and-a-half times its body length, longer than any other mammal. This bat seems to have evolved its incredible tongue to feed on a tubular flower found in the cloud forests of Ecuador. It mops up the flower's nectar with tiny hairs on the end of its tongue. It also collects some pollen on its head, which it spreads from flower to flower, helping the plant to create seeds.

Butterflies and moths eat with a proboscis, which is curled up like a coiled tape measure and unfurls to extend into a flower's centre. It may look like a long, thin straw, but it works more like a paper towel that absorbs the flower's sweet nectar.

A BUTTERFLY WITH ITS PROBOSCIS CURLED

CONTESTANT #4

CONTESTANT #5

TONGUE STORAGE

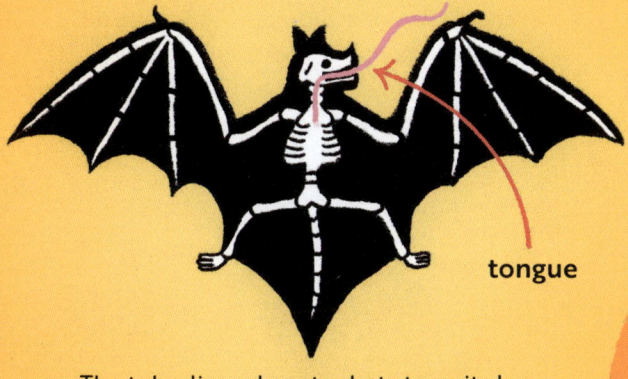

tongue

The tube-lipped nectar bat stows its long tongue inside its ribcage.

THE WALLACE'S SPHINX MOTH

Naturalists Charles Darwin and Alfred Russel Wallace separately predicted the existence of a moth with a super-long proboscis more than 40 years before it was discovered. This was because they came across a species of orchid with nectar in a very hard-to-reach spot far inside the flower and thought there should be an insect with a really long proboscis to match. This insect is Wallace's sphinx moth (also known as Darwin's moth), which lives in Madagascar and has a tongue that is more than three times longer than its body!

The winner is.... the Wallace's sphinx moth! With a 20–30-centimetre-long proboscis, this animal has the longest tongue (relative to body size) in the world.

SURPRISING SPIT

Saliva has superpowers: it helps us to enjoy and digest food, fight nasty germs and talk. It's made in special pouches called salivary glands, which are dotted all around our mouths. When you see, hear or smell something that tickles your appetite, your brain commands these glands to produce a lot of saliva. We produce between 0.5 and 1.5 litres of saliva every day, which is between two and six glasses. But this is nothing compared to cows, which make between 40 and 190 litres of saliva a day!

Prairie dogs will often jump up on their hind legs and make a 'yip' noise. This action tells predators that the colony is alert. It also helps the colony bond.

ANTS SWAPPING SPIT

Have you been told not to talk with food in your mouth? Social insects, such as ants, bees and wasps, often walk around with their mouths full. They pass food to one another through mouth-to-mouth exchanges. Along with food, ants transfer spit, which contains chemicals that help them grow, digest food and defend themselves against diseases.

KISSING FRIENDS OR ENEMIES

When prairie dogs kiss, it can be a friendly greeting or the start of a territorial fight. The spit exchange involved in kissing helps them recognise each other or say 'hi'. For example, a mother kisses a wandering baby to check if it's hers, and adults kiss mouth-to-mouth when they meet. However, if they realise that they belong to two different groups, they jump away from the kiss and squeak. The prairie dog of that territory will then try to chase away the intruder.

HOW SILKWORMS MAKE COCOONS

1 A silkworm spends its first month or so of life feeding on the leaves of mulberry trees. These leaves provide all the nutrients the silkworm needs.

2 When it is about one to two months old, the silkworm stops eating and begins to form a net to hold its cocoon in place. Then, it weaves the cocoon from the outside in by moving its head in a figure-of-eight motion.

3 After about three days of spinning, the silkworm is completely enclosed in its cocoon.

SILKY SPIT

Silkworms aren't worms at all – they're caterpillars and they are native to China. They spin cocoons around themselves to allow their bodies to go through an incredible change. A silkworm will release a trail of spit from its mouth, which solidifies into thread. The thread is now called silk, and silkworms wrap it around themselves to form the cocoon. Humans use the silk to make fabric, so next time you have a chance to feel something made of silk, remember it's just dried spit!

4 A silkworm cocoon is made from a single strand of silk: it can be 800 metres long, or even longer. The cocoon is like a sleeping bag that protects the silkworm. However, the silkworm is not sleeping, it's transforming...

Can you guess what will emerge from the cocoon?

The silk moth! Say, 'Hi!'

A FIVE-DAY LIFE

Inside the cocoon, the silkworm sheds its skin, develops wings and turns into a white, hairy moth. This process is called metamorphosis, and it takes between two and three weeks. Silk moths live for around five days and never eat: for energy, they rely on the leaves they ate as a silkworm before forming the cocoon.

The moth escapes from the cocoon by producing a fluid that dissolves the silk, making a hole.

Silk moths usually emerge from their cocoons early in the morning. They stretch their crumpled wings but cannot fly. They mate and lay eggs.

SPIT BATHS

Just as humans can cool down by getting their skin wet, kangaroos try to beat the heat by licking themselves. They give their arms a refreshing 'spit bath', which helps the network of blood vessels in their forearms cool down their whole body. Their temperature lowers as the saliva evaporates.

SNEAKY SPIT

Female mosquitoes need to drink blood so their eggs will develop. Their special saliva helps them drink an animal's blood without the animal even noticing. That's sneaky, but how does it work? Mosquito spit contains several chemicals that keep the animal's blood flowing, make the blood vessels a bit wider and even make the skin almost numb. Only later does the spot becomes red, itchy and swollen. The needle-like parts of the mosquito's mouth that pierce the skin form the fascicle, while the labium (lower lip) retracts and stays outside the skin.

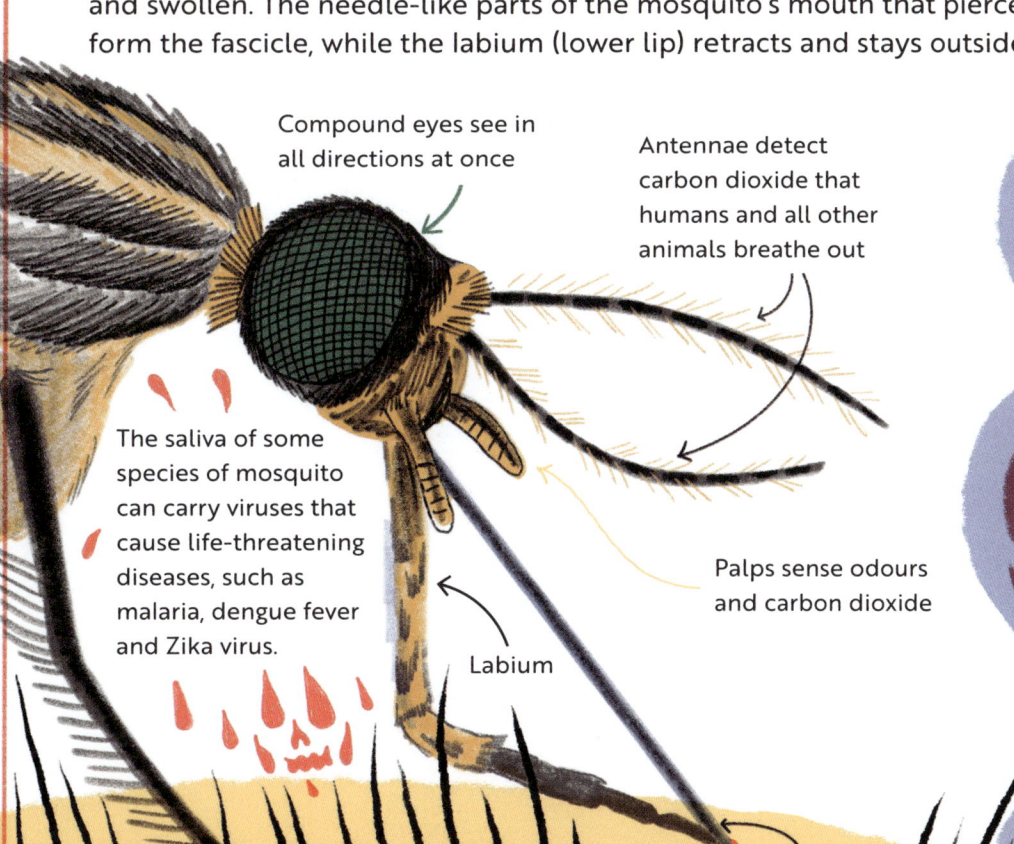

Compound eyes see in all directions at once

Antennae detect carbon dioxide that humans and all other animals breathe out

The saliva of some species of mosquito can carry viruses that cause life-threatening diseases, such as malaria, dengue fever and Zika virus.

Palps sense odours and carbon dioxide

Labium

Fascicle

SALIVA NESTS

Cave swiftlets make white or yellowish nests with their hardened saliva. The nests look like little hammocks of tightly woven threads.

BILLS & BEAKS

Birds' beaks come in a wide range of shapes suited for catching and eating different kinds of food. Beaks can be large or thin, straight or curved. They can work as weapons or combs to preen feathers or might just be a great way to show off. Can you identify any of these birds just from their beaks?

GUESS WHO?
A short, curved beak with a sharp tip can be useful for splitting open hard fruits and nuts. Can you guess the name of this talkative bird?

GUESS WHO?
This colourful bill is perfectly designed for carrying fish. It also glows under ultraviolet light, but scientists don't yet know why.

GUESS WHO?
A powerful, deeply hooked beak is a typical feature of birds of prey. This one has very sharp cutting edges to tear and slice through skin and flesh.

GUESS WHO?
On top of this fancy beak, there is a horn made of keratin. This horn is called a casque and it looks a bit like a rhinoceros horn. This bill is good for eating fruits.

THE MANY USES OF KERATIN
Beaks are covered with keratin, the tough, flexible protein that also makes up human fingernails, porcupine quills and cow horns, as well as feathers and hair.

GUESS WHO?
This bird's beak can reach 45 centimetres in length, and has a super elastic throat pouch that can scoop up around 11 litres of water in one go.

GUESS WHO?
This flightless bird lives in New Zealand. It keeps its long, narrow beak to the ground, sniffing and probing for tasty worms.

GUESS WHO?
The beak of this animal can be hidden by a long, dangling bit of flesh, called a snood. The snood can change colour, size and shape according to the bird's emotions.

GUESS WHO?
This bird lives in the hot regions of central and eastern South America. Its large beak helps it stay cool. Can you guess how?

GUESS WHO?
Shaped a bit like a banana, this curved bill belongs to a tall, pink bird that looks for food in ponds and riverbeds. It uses its beak to sift food from the mud and water.

Are you a master bird watcher? Turn the page to find out.

Take a look at these birds to find the answers. Now you're the expert, try quizzing your grown-ups!

PARROT
Parrots are probably one of the most intelligent bird species. Some can mimic humans and repeat words with the help of their unique tongues.

PUFFIN
Puffins are famous for their special beak. Its colours become brighter during the breeding season. A colourful beak may be a sign of good health, making the puffin that has one a more attractive mate.

BALD EAGLE
Bald eagles are skilled hunters of fish, other birds and small mammals.

RHINOCEROS HORNBILL
Rhinoceros hornbills are found in Southeast Asia. Their outstanding casque is hollow and is probably used as a megaphone to make loud calls during courting rituals and fights.

WHY SO MANY DIFFERENT BEAKS?

In 1835, the famous naturalist Charles Darwin journeyed around the Galápagos Islands in South America. He found that there were more than 12 distinct species of finch, all with different beaks. It turned out that each finch had adapted to eat different food, from big seeds and small seeds, to fruits, cacti or insects.

KIWI
The kiwi has a nose at the end of its beak. As it walks, it taps its beak on the ground to detect smells and vibrations. When it detects prey, the kiwi pushes its beak into the soil. To get its beak as deep as possible, it can kick its legs in the air, putting its entire weight downwards in a 'beakstand'.

PELICAN
Brown pelicans dive into the water at speeds of around 64 kilometres per hour to catch prey. Their eyesight is so sharp that they can spot a fish in the ocean while flying 20 metres above the water's surface.

TURKEY
Female turkeys prefer males with the longest snoods above their beaks. The snood's colour is important too: a bright red snood shows that the turkey is excited or aggressive, while a pale snood is a sign of sickness.

TOUCAN
Toucans cannot sweat, but their bright, light and long beaks help them to cool off. When it is too hot, blood flows to the beak and releases heat into the air.

FLAMINGO
This is a flamingo. These birds are famous for their pink colour, slender legs and long necks. They become pink by eating a lot of algae and shrimp, which contain carotenoid (the same substance that makes carrots orange).

HIDDEN TALENTS

In some action films, spies and criminals hide poison or secret messages inside false, hollow teeth. Quite a clever trick, don't you think? Well, humans aren't the only animals that use their mouths in ingenious ways.

FORGOT YOUR SHOPPING BAG?
Is this hamster biting off more than it can chew? No, it's stuffing its cheek pouches so it can save the food for later. It uses the stretchy pouches like shopping bags to carry a lot of food to another place, keeping it safe and dry. Chipmunks, platypuses and some monkeys have cheek pouches, too.

Hamsters' stretchy cheek pouches extend all the way to their hips and can expand to nearly triple their original size when full.

WHAT'S THIS LION DOING?
Roaring? Snarling? Yawning? No, this lion is sniffing the air through his open mouth. This weird grimace is called a flehmen response. The German word *flehmen* means 'to curl the upper lip'. But what's this lion sniffing? Just as dogs pee on every other tree, wild animals leave their smelly markings wherever they like. An organ on the roof of their mouths allows lions to understand whether a smell was left by a rival or a potential mate. Many other mammals, such as horses, goats and deer, show the flehmen response, too.

UNDERWATER EXCAVATORS

Yellowhead jawfish live around the edges of reefs. They dig burrows by stuffing their mouths with sand and then spitting the sand out onto the sea floor. Even after a burrow is built, they spend a lot of time keeping it clean by removing sand and stones that drift in.

At night-time, a yellowhead jawfish will hide in its burrow and cover the opening with a rock or shell.

If an unwanted visitor comes near a yellowhead jawfish's burrow, it might open its mouth wide, showing off its big, deep throat to scare the intruder away.

When a yellowhead jawfish is scared, it quickly drops into its burrow tail first.

All of these jawfish except the one in the middle seem very busy. What's the middle fish doing?

He's protecting his eggs — by holding them in his mouth!

WHAT A MOUTHFUL!

Male yellowhead jawfish carry their eggs in their mouths. The eggs also hatch inside the dad's mouth. The big advantage of this scheme is that the dad can escape from dangerous predators without abandoning his young.

The dad releases his offspring (called fry) after seven to nine days, when they are about 4 millimetres long and ready to fend for themselves.

Having a nursery inside your mouth has a downside. The dad can't eat, otherwise he would risk losing or swallowing the fry. The yellowhead jawfish isn't the only mouthbrooding father in the animal kingdom. Others include some types of cardinalfish and the hardhead catfish.

To provide the eggs with some oxygen and remove waste, the dad spits out all the eggs and then quickly sucks them back in.

COLOURFUL MOUTH PRINTS

Every crested coua chick is born with unique, colourful markings inside its mouth. These red-and-white decorations make it easier for the parents to see the chick's open mouth in a dark nest. It may also help the parents tell their chicks apart or show them where to place food. The markings fade over time.

These chicks develop other colourful features as they grow older: long purplish-blue feathers, a white belly, a pointed crest, a black bill and beautiful turquoise skin around their eyes.

NEED A RIDE?

Some types of remora fish enjoy a free ride around the ocean by clinging to whale sharks using the suckers on top of their heads. They usually attach themselves to a whale shark's belly or side, or inside its mouth.

This may seem dangerous for the remora fish, but there's no need to worry: whale sharks eat mostly plankton, tiny fish and squid, so the remora are safe. During the trip, the remora pay for their ride by cleaning scraps and parasites off the shark and getting a free meal in the process!

MIGHTY MOUTHS

We can use our mouths for laughing, smiling, speaking, singing, whistling, blowing, kissing, yawning and many other things besides. Some people who cannot use their hands or feet even learn to paint and write with their mouths. So far, we have met animals who use their mouths to store food, hatch eggs, give rides, smell, dig, spin cocoons, build nests, cut down trees, recognise friends, and of course, hunt and eat. What else can animals do with their mouths? Lots, as it happens.

A goby can climb slippery rocks thanks to a sucker in its mouth and another on its belly. But before it can start its climb, it needs to make a dramatic change – moving its mouth from the front of its body to underneath! This transformation takes about two days.

CLIMBING WATERFALLS

Rock-climbing goby fish called 'o'opu nōpili are found throughout Hawaii. They feed by scraping tiny bits of algae off the surface of rocks with their mouths. They hatch in streams but are washed to the ocean by the currents. After several months, they migrate back, going against strong currents and climbing waterfalls as tall as 100 metres – using their mouths!

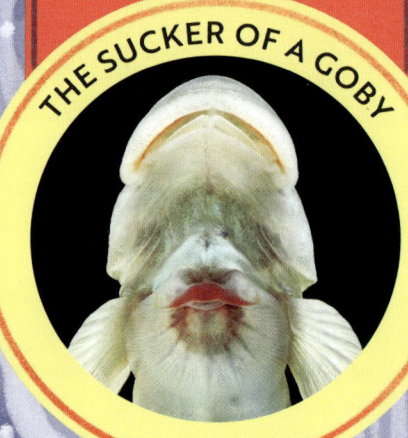

THE SUCKER OF A GOBY

LIP WRESTLING

The biggest mouth wins! Many species of cichlid grab each other by the lips to wrestle. Cichlids also press their lips together as a sign of affection, but this behaviour is nowhere near as common as the aggressive jaw-locking.

Not for the Squeamish

Jellyfish both eat and poo using the same small opening! They have stinging tentacles, which are perfect for stunning their prey and pushing it towards their mouths, in the centre of their bodies. They gobble up fish, shrimps, crabs and plants.

SHOWING HOW YOU FEEL

The mouth can show mood and emotions. How do your mouth and face change when you are concentrating, angry, happy, confused, surprised, sad, afraid or in pain? Other animals also make expressive faces, but not always in the same way we do. For example, when we pull back our lips to bare our teeth, it's usually a friendly smile. But when growling dogs bare their teeth, they are saying 'Back off now!' Barbary macaques have many things in common with humans, but their facial expressions might not mean what you think...

A The eyebrows of this Barbary macaque are raised and its mouth is open showing its teeth.

B This macaque has its mouth closed and a wrinkled brow.

C The corners of this macaque's lips are fully drawn back, showing its upper and lower teeth.

D This macaque has raised eyebrows and a mouth shaped like an O.

E The mouth of this macaque is wide open, like a big yawn.

Can you work out how these Barbary macaques are feeling based on their facial expressions?

55

BUILDING AND STITCHING

Most birds rely on their beaks and feet to collect and transport all the construction materials they need to craft their home sweet homes. Their nests must be strong and safe to protect and raise their hatchlings until they are ready to fly. They must also be sturdy enough to withstand bad weather conditions. Here are two very intriguing nest masterpieces.

SOCIABLE WEAVER BIRD

The sociable weaver bird's nest is among the largest built by any bird. It is like a building with multiple flats in it, each occupied by a pair of birds. These colony builders live in South Africa, Namibia and Botswana. They carry twigs, straw and grasses with their blunt, cone-shaped bills, which are also ideal for feasting on seeds, grain and insects. Their homes provide shade from direct sunlight, as well as protection from rain and cold.

COMMON TAILORBIRD

Found across southern Asia, common tailorbirds are small songbirds that stitch their nests with their beaks. The female carefully chooses a strong but soft leaf. She then pierces many tiny holes along the leaf's edge, using her long, slender, beak like a needle. The bird threads plant fibres or insect silk through the holes to close up the nest, and ta-da! A cosy space for a hatchling.

SCOOPING AND SHOVELLING

Naked mole rats have just four teeth, but they are useful tools. They use them as shovels to dig their burrows, as weapons and, of course, for eating. Naked mole rats can even move their lower incisors independently, like chopsticks, and can close their lips behind their teeth when they dig to keep dirt out of their mouths.

GIGANTIC GRINS

The bowhead whale holds the Guinness World Record for the most colossal mouth in the world. A bowhead whale's mouth can be 5 metres long, 4 metres high and 2.5 metres wide. But, these huge marine animals feed mainly on some of the ocean's tiniest creatures. Their diet is made up of a group of organisms called zooplankton, which include krill and small crustaceans called copepods. In order to satisfy its appetite, a bowhead whale needs to eat about 2 tonnes of these little organisms every day.

Adult female bowhead whales can weigh around 90 tonnes and measure 20 metres in length. Newborn bowhead whales measure around 4 metres long and weigh 'only' about a tonne, which is more than ten human adults!

Bowhead whales can live for a really long time: more than 200 years! Killer whales (orcas) are their only known natural predator.

Scientists believe that baleen whales sing to communicate with each other underwater. Every year the whales change their tunes, but scientists don't yet know why. Their songs can travel thousands of kilometres, but can be disrupted by human-made noises from ships, oil-rigs and other undersea activities.

These whales use their big heads, which are about one-third the length of their bodies, to break through thick ice.

One of the bowhead whale's distinctive features is its white chin, decorated with black spots.

Such a big mouth must have a huge number of massive teeth, right? Guess how many.

Zero! These whales gulp tonnes of water at a time and use a broom-like structure called a baleen plate as a sieve to strain out their prey. No teeth required!

Baleen is made from keratin, the same stuff that makes up our hair and nails. It grows down from the whale's top jaw.

SPOT THE DIFFERENCE
TOOTHED vs BALEEN WHALE

Whales are divided into two categories – toothed whales, which include dolphins, killer whales, narwhals, sperm whales, pilot whales and beluga whales, and baleen whales, which include humpback and bowhead whales. You can see the difference as soon as they open their mouths.

TOOTHED WHALE

BALEEN WHALE

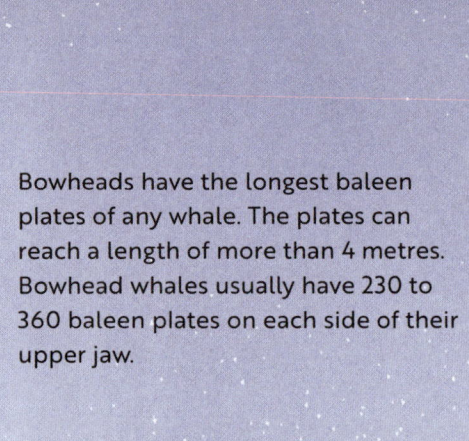

Bowheads have the longest baleen plates of any whale. The plates can reach a length of more than 4 metres. Bowhead whales usually have 230 to 360 baleen plates on each side of their upper jaw.

KRILL UP CLOSE

Zooplankton, such as krill, are only a few millimetres long. Despite their small size, they are important for the health of the ocean ecosystem. Climate change and overfishing threaten the krill population and put the whales' lives in danger.

As a bowhead whale swims towards a swarm of zooplankton with its mouth partly open, it takes a big gulp of water. It then pushes the water back out of its mouth through its baleen. The krill, small crustaceans and other zooplankton become trapped by the baleen. The whale can now swallow its captured prey.

SAY 'CHEESE'!

From the wildebeest in the African savanna to the bowhead whales in the Arctic, now we know how many extraordinary animals chomp, construct, carry, climb and so much more, with just their mouths! We've met some of the biggest, smallest, fiercest and stickiest mouths and beaks in the animal kingdom – which is your favourite? Has the way that you think about your own marvellous mouth changed now that you know just how awesome it is? It's time to smile and show off those pearly whites. Let's celebrate one of the most important parts of our, and all animals', bodies – our crunching, snapping, slurping and all-round magnificent mouths.

GLOSSARY

baleen
A large, flat structure made of keratin, that hangs down from the upper jaws of baleen whales. The whales use it to filter the seawater for food.

browser
An animal that eats mainly leaves, fruits, soft shoots and shrubs.

canine tooth
Pointed tooth that's usually in the front of the mouth and that is useful for grasping and tearing food.

carnassial
A large tooth that meat-eating animals have and that is suitable for cutting flesh. Carnassials are usually found in the upper and lower jaws, and they work together like scissors.

carnivore
A meat-eater.

cementum
A bone-like material in mammals' teeth that helps keep the tooth in the gum.

crustacean
A type of animal that doesn't have a backbone but does have a hard shell, such as a crab, lobster or woodlouse.

denticle
A microscopic tooth-like structure.

dentine
A hard substance that makes up the bulk of teeth.

enamel
The substance that forms the outer surface of mammals' teeth.

food chain
A chain of living things that eat other living things. For example, plants are eaten by insects; insects are eaten by small birds, such as sparrows; and sparrows are eaten by peregrine falcons.

grazer
An animal that eats mainly grass.

herbivore
A plant-eater.

incisor
A front tooth that is useful for tearing, slicing and holding things.

invertebrate
An animal without a backbone.

keratin
A tough protein that covers horns and makes parts of animals' bodies such as nails, hair and baleen. Keratin is also found in birds' beaks and feathers, and reptiles' claws and shells.

larva
A baby insect.

mammal
An animal that has a backbone and can regulate its own body temperature. Most mammals have fur or hair. Baby mammals drink milk that their mothers make in their bodies.

mandible (in insects)
Jaws that move sideways and are used for gripping, biting and cutting.

melanin
A dark substance that gives colour to the skin, hair and eyes.

metamorphosis
A transformation – for example, a caterpillar turning into a butterfly.

mimic
To copy or imitate.

molar
A grinding tooth at the back of the mouth.

mollusc
A type of animal that is soft and squishy with a hard shell, such as a snail or a clam.

mouthbrooding
A strategy for looking after young where a parent keeps eggs (and sometimes babies) inside their mouth for protection.

naturalist
A person who loves and studies nature.

nymph
A baby insect that needs to go through a transition to grow and look like its parents.

offspring
A baby animal.

organism
A living thing.

papillae
Small structures found on the tongue that help with tasting food, sensing texture and detecting chemicals. The papillae on your tongue look like little bumps.

plankton
Tiny creatures that live in the sea but cannot swim. Instead, they are carried along by tides and currents.

predator
A meat-eater that feeds on other animals.

premolar
A type of tooth that is used for grinding and that is found in between the canines and molars.

prey
An animal that gets eaten by a predator.

proboscis (in insects)
A long mouthpart used for sucking nectar or for piercing a victim's skin and drinking its blood.

radula
A tongue-like structure that is covered in tiny tooth-like denticles that some animals, such as snails, use to eat.

regurgitate
To bring food back up into the mouth after eating and swallowing it.

savanna
A grassland with few trees found in tropical areas of the world.

scavenger
An animal that does not hunt, but instead feeds on dead animals or other food that it finds.

species
A big family of animals that look similar and can have babies together.

territory
An area that belongs to an animal and they do not want other animals to enter.

ultraviolet light
A type of light that humans can't see but some animals can.

venomous
An animal that has a harmful substance, called venom, in its body that it can inject into other animals with a sting or a bite.

vertebrate
Any animal with a backbone, such as fish, amphibians, reptiles, birds and mammals, including humans.

INDEX

A
adult teeth 6, 14
alligators 14, 22
ambushes 11
anacondas 23–4
anteaters 38
ants 42
apex predators 11
Argentinosaurus 25

B
bacteria 6, 9
baleen plates 60, 61
bats, tube-lipped nectar 41
beaks and bills 9, 10, 46–9
bears 7, 31, 35, 36
beavers 30, 31–2
birds of prey 10
blood, drinking 4, 45
browsing 18–19, 21
burps 19
burrowing 51, 57

C
caimans 22, 23–4
camels 18
canines 6, 7, 12, 29
carnassials 7, 12
carnivores 7
cats 7, 37
cementum 6, 27
chameleons 39–40
chipmunks 50
cichlids 54
claws 17
cleaning teeth 6
climbing 4, 54
cocoons 43–4
couas, crested 53
cows 19, 34
crocodiles 14, 22

D
deer, water 29
denticles 8
dentine 6, 15, 28
dinosaurs 17, 25
dogs 7, 50
dragonflies 13
drinking 37

E
eagles, bald 46, 48
eggs 52
elephants 27–8
enamel 6, 15, 28

F
facial expressions 55–6
falcons 10
fangs 24
flamingos 47, 49
flehmen response 50
fluoride 16
food chain 11
food storage 4, 50
fossils 17
frogs 25

G
giraffes 7, 35, 36
gizzards 22
gnawing 30–2
go-away birds 4
gobies, Nopili rock-climbing 54
grazers 18–19
grinding pads 9
grit 22
grooming 37
gums 6

H
hamsters 50
hawks 10
herbivores 7, 18
hippopotamuses 29
horses 19–20
humans 6

I
incisors 6, 7, 12, 19, 27, 33
iron 32
ivory 27

J
jawfish, yellowhead 51–2
jellyfish 54

K
kangaroos 45
kiwis 47, 49
Komodo dragons 35, 36

L
licking 34, 37
lions 4, 50
lizards 35, 36
louse, tongue-eating 36

M
macaques, Barbary 55–6
mammals 7
mandibles 13
marmots 33
mastodons 28
milk teeth 6
molars 6, 7, 12
mole rats, naked 57
molluscs 7, 8
mosquitoes 4, 45
moths, Wallace's sphinx 41

NO
narwhals 26
nests 45, 57
oxpeckers 21

PQ
papillae 18, 34
parrots 46, 48
pelicans 47, 49
penguins 22
platypuses 9, 50
polar bears 35, 36
poo 17, 33, 54
porcupines 31, 32, 33
pouches 50
prairie dogs 42
prasinohaemas 35, 36
praying mantises 13
premolars 6, 7, 12
puffins 46, 48
pulp 6
quills 32

R
rabbits 30, 33
radula 8
rats 30
regurgitation 22
remora fish 53
rhinoceroses 21
rodents 30–2
roots, teeth 6

S
saliva 42–5
sea sponges 9
sharks 14, 15–16, 53
sheep 7
silkworms 43–4
skinks, blue-tongued 35
snails 7, 8
snakes 23–4
spit 42–5
squirrels 30, 31
swallowing whole 22–5
swiftlets, cave 45

TV
T. rex 17
tasting 37
tigers 11–12
tongues 25, 34–41
toothless animals 9
toucans 47, 49
turkeys 47, 49
tusks 26–9
venom 24
vomit 30

WZ
walruses 26
weasels 7
weaverbirds, sociable 57
whales 58–61
wildebeest 4
wolves 7
woodpeckers 38
woolly mammoths 28
worms, giant tube 9
zooplankton 58, 61

SELECTED SOURCES

Here is a selection of the books and articles that the author used as reference material for this book.

Pages 6–9 Pearly Whites
'Everything you need to know about teeth' www.nhsinform.scot
'Cold-seep Tubeworms' https://oceanexplorer.noaa.gov/explorations/02mexico/background/tubeworms/tubeworms.html

Pages 10–13 Attack!
'Praying Mantis' www.kids.nationalgeographic.com
'Tiger guide: species facts, how they hunt and where to see in the wild' www.discoverwildlife.com

Pages 14–17 Losing Teeth
'How Many Teeth Do Sharks Have and Other Sharks' Teeth Facts' www.discoveryuk.com/sharks/how-many-teeth-do-sharks-have-and-other-sharks-teeth-facts
'What's the Difference Between Alligators and Crocodiles?' www.britannica.com

Pages 18–21 Chew it Over
'Ruminant' www.britannica.com
'This is how camels eat spikey cacti' www.nationalgeographic.com

Pages 22–25 Down in One
'This is how a frog's tongue works' www.australiangeographic.com.au/news/2017/02/this-is-how-a-frogs-tongue-works-2/
'What is the biggest snake in the world?' www.nhm.ac.uk

Pages 26–29 Tremendous Tusks
'What is a narwhal?' https://oceanexplorer.noaa.gov/facts/narwhal.html
'The History of the Ivory Trade' www.nationalgeographic.org/society

Pages 30–33 Grow & Gnaw
The Encyclopedia of Animals: A Complete Visual Guide. Fred Cooke et al, 2008.

Pages 34–37 Terrific Tongues
'Fact or Fiction?: The Tongue Is the Strongest Muscle in the Body' www.scientificamerican.com/article/fact-or-fiction-the-tongue-is-the-strongest-muscle-in-the-body
'The surprising physics of cats' drinking' https://news.mit.edu/2010/cat-lapping-1112

Pages 38–41 Longest Tongue
'Xanthopan morganii praedicta' www.britannica.com
'Chameleon spit is 400 times thicker than human's' www.science.org/

Pages 42–45 Surprising Spit
'How Is Silk Made? The Ethical Dilemma of Its Origins' www.discovermagazine.com/planet-earth/silk-making-is-an-ancient-practice-that-presents-an-ethical-dilemma
'The Roo Files' www.pbs.org/wnet/nature/big-red-roos-the-roo-files/2415/

Pages 46–49 Bills & Beaks
www.rspb.org.uk
'Evolution teaching resource: spot the adaptations in Darwin's finches' www.nhm.ac.uk/

Pages 50–53 Hidden Talents
'Golden Hamsters' Amazing Cheek Pouches' https://www.bbc.co.uk/programmes/p02gxlt8'Daddy Day Care Underwater' https://blogs.scientificamerican.com/observations/daddy-day-care-underwater/

Pages 54–57 Mighty Mouths
'Experience-based human perception of facial expressions in Barbary macaques (Macaca sylvanus)'. Laëtitia Maréchal, Xandria Levy, Kerstin Meints, Bonaventura Majolo, 2017.
'10 Things You Didn't Know About Naked Mole-rats' https://nationalzoo.si.edu

Pages 58–61 Gigantic Grins
www.guinnessworldrecords.com/world-records/largest-mouth
'bowhead right whale' www.britannica.com/

PICTURE CREDITS

We would like to thank the following for their kind permission to use their photographs:

Page 15. The dermal denticles of a lemon shark, viewed through a scanning electron microscope. Pascal Deynat/Odontobase.

Page 16. Grey nurse shark. Joe Lencioni.

Page 24. Snake Attack! Matthijs Kuijpers/Alamy.

Page 26. Narwhal (Monodon monoceros) male surfacing, Baffin Island, Nunavut, Canada. Flip Nicklin/Minden Pictures/Alamy.

Page 37. Cat tongue. Patricia Vottero/iStock.com.

Page 41. Butterfly tongue or its correct name is a proboscis. Russell Marshall/Shutterstock.

Page 42. Trophallaxis in black ants (Camponotus cf. compressus). Rakesh Kumar Dogra, IFS Indian Forest Service, Tamil Nadu Forest Department.

Page 54. Hildebrand's goby (Sicydium hildebrandi) sucker fish from Dagua River basin, Colombia. Dan Olsen/Shutterstock.

Page 61. Northern krill (Meganyctiphanes norvegica). Øystein Paulsen.

A note of thanks from the author

I'm so grateful to all the amazing people who supported this book. Hats off to the What on Earth Publishing team for their hard work and dedication. In particular, I would like to thank Patrick, Andy, Nancy, Laura and Katy, who turned my initial idea into this dazzling product. I'd also like to thank Richard for checking all the facts.

I give a cheerful, virtual high five to Ed J. Brown: his bright colours and captivating compositions have transformed this book into a feast for the eyes.

I would like to say thank you to Dr Laëtitia Maréchal for providing information about macaques and Dr Kelly Diamond for information on rock-climbing fishes. Their enthusiasm and help was much appreciated.

Warm hugs go to all my friends and family. You are my ultimate source of joy, creativity and inspiration! A very special thanks to my niece Giorgia, who cannot stop learning about animals. Soon she will be able to read and I cannot wait to read this book together with her.

And I'd like to say a thank you as big as a bowhead whale's mouth to you, the reader! I hope you love this book and that it inspires you to learn more about the animal kingdom. There's still so much to discover about animal teeth, spit and tongues – more than could fit into these pages. So, keep on asking questions and exploring the wonders of nature.

Letizia

www.letiziadiamante.com